Gisbert Kapp

Transformers for Single and Multiphase Currents

A Treatise on their Theory, Construction, and Use

Gisbert Kapp

Transformers for Single and Multiphase Currents
A Treatise on their Theory, Construction, and Use

ISBN/EAN: 9783337185893

Printed in Europe, USA, Canada, Australia, Japan

Cover: Foto ©berggeist007 / pixelio.de

More available books at **www.hansebooks.com**

TRANSFORMERS

FOR

SINGLE AND MULTIPHASE CURRENTS

A TREATISE ON

THEIR THEORY, CONSTRUCTION, AND USE

BY

GISBERT KAPP

MEMB. INST. C. E., MEMB. INST. E. E.

Translated from the German by the Author

WITH ONE HUNDRED AND THIRTY-THREE ILLUSTRATIONS

WHITTAKER AND CO.

WHITE HART STREET, PATERNOSTER SQUARE, LONDON

AND 66, FIFTH AVENUE, NEW YORK

1896

PREFACE

In what may be termed the literature of heavy electrical engineering there is no lack of books on dynamos, motors, cables, and a host of auxiliary apparatus, but transformers have been somewhat neglected. The reason may possibly be that scientists have not considered it worth their trouble to investigate so seemingly simple a piece of apparatus as a transformer, whilst so much more interesting problems connected with machinery in motion remained to be solved. Practical engineers, on the other hand, who, in carrying on their profession, must investigate whatever they design, cannot be expected to publish the result of their researches for the benefit of competitors. Notwithstanding the apparent simplicity of the transformer, a study of this apparatus leads to a number of questions which are not only highly interesting from a purely scientific point of view, but are also very important for practical or commercial reasons. On most alternating-current lighting installations the transformers are at work night and day all the year round, and cause a waste of power which is always going on. Under these circumstances, even a small improvement in the efficiency of the apparatus has a large monetary value, but such improvements can only be made as the result of careful study combined with practical experience.

The object of the present book is to enable the reader to judge the design of transformers and to design such apparatus for himself. The mathematical treatment of the subject has been kept as short and as simple as possible. I am well aware that through the omission of many theoretical considerations which were not immediately pertinent to the practical object I had in view, my book, considered as a scientific treatise on transformers, is incomplete. I have omitted to include detailed researches on the influence of the shape of current and E.M.F. waves on hysteresis loss; I also have omitted the analysis of such waves by Fourier's series, and generally the analytical treatment of problems connected with transformers. These are sins of omission for which I plead the indulgence of the reader, on the ground that my book is not intended for mathematicians or physicists, but for engineering students and practical engineers, whose object must always be to obtain a maximum of practical success with a minimum expenditure of mental labour.

GISBERT KAPP.

Berlin, September 1896.

CONTENTS

CHAPTER I

CHAPTER II

CHAPTER III

CHAPTER IV

CHAPTER V

CONTENTS.

CHAPTER VI

CHAPTER VII

CHAPTER VIII

CHAPTER IX

LIST OF ILLUSTRATIONS

TRANSFORMERS

CHAPTER I

PRINCIPLE OF ACTION—MAGNETIC LEAKAGE—ARRANGEMENT OF COILS
—FUNDAMENTAL EQUATION

Principle of action.—If the magnetic flux N passing through a coil changes, an E.M.F. is induced in the coil which is proportional to the time-rate of change $\left(\dfrac{dN}{dt}\right)$ and the number of turns n. Conversely, if a current be sent through the coil it produces a magnetic flux, threading the coil, which is within certain limits proportional to the current. If this current changes, a corresponding change takes place in the magnetic flux. Let now two coils be so arranged that the flux produced by the current in one passes wholly or partially through the other, then any change in the current strength in the former coil will produce an E.M.F. in the latter coil. Such an arrangement is shown in Fig. 1, where a ring of iron is threaded through the two coils I, II. A current passing through coil I produces a magnetic field which passes partly through the iron ring, and partly through the air surrounding this coil. The flux will therefore be strongest in the centre of the coil, at a, and weakest at b, in the centre of coil II.

B

The iron ring acts as a vehicle for carrying the flux produced by coil I through coil II, though as an imperfect vehicle, since part of the flux is lost on the way. In a sense, the iron ring may be regarded as a magnetic link between the two coils. Even without iron the coils can be linked together by the magnetic flux passing through air. Thus in the position shown, the field produced by I would in part pass through II, though its strength would be much diminished. The same holds good if the two coils are laid upon each other, in which position a somewhat stronger field would pass through II, though not so strong as with an iron ring. If, however, whilst

Fig. 1.

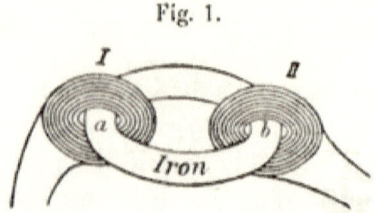

still omitting the iron ring, the coils are relatively so placed that the axis of I lies in the plane of II, or *vice versâ*, then none of the lines produced by I can pass through II, and a change in the current passing through I cannot produce any E.M.F. in II. By suitably placing the coils, an inductive effect of one upon the other can therefore be produced, even without the use of an iron link, but the employment of such a link has the advantage that not only is the inductive action increased, but it becomes to a greater extent independent of the mutual position of the coils. An apparatus consisting of two coils, interlinked with an iron coil common to both, is called a transformer.

It has already been mentioned that the E.M.F. produced in II, which we may call the secondary coil, is proportional to the time-rate of change of the current in the primary coil I. Since the current in this coil cannot alter indefinitely in the same sense without becoming infinite, it follows that periods of growing current strength must alternate with periods of declining current strength. If, then, with a growing current in the primary coil the E.M.F. induced in the secondary coil acts in one way, it must act in the opposite way if the current diminishes, and it is thus clear that changes in the current strength in the primary coil, even if not accompanied by changes in direction, must produce an alternating E.M.F. in the secondary coil. This alternating E.M.F. produces, in an external circuit connected to the terminals of the secondary coil, an alternating current. We are thus able to convert a unidirected pulsatory current into an alternating current, but never into a continuous current. Instead of using a pulsating current in the primary coil, we may with advantage use an alternating current, and thus obtain from the secondary coil another alternating current, the E.M.F. of which is dependent on that of the primary current, and on the ratio between the number of turns in the two windings.

Magnetic leakage.—Since the lines of force not only pass through iron, but in a lesser degree also through air, it follows that only part of the magnetic flux in a actually threads through the secondary coil at b, the rest closing round the primary coil in air. The difference between the flux in a and b will be the greater the farther the coils are from each other, and the greater the resistance which the iron offers to the passage of the lines of force. In consequence of this resistance (sometimes also called magnetic reluctance) lines of force are caused to leak out laterally, and form thus a leakage field which does not contribute in any way to the production of E.M.F. in the secondary coil. The more leakage there is, the smaller is consequently the E.M.F. induced in the secondary coil through an alternating current in the primary coil.

In order to understand the conditions which influence leakage, we assume for the present that the primary coil carries a continuous current, whilst through the secondary coil there passes, either no current at all, or also a continuous current of such direction as will tend to weaken the field produced by the primary current.

The coil I drives, then, a magnetic flux in a certain direction through the iron ring. If no current flows through coil II, then the lines of force have only to overcome the magnetic resistance of the iron path, which may be so small that comparatively few lines are crowded out. If, however, the coil II also carries a current, it will tend to produce a magnetic flux in the opposite direction, which, colliding with the original flux, must cause a strong leakage field, thus weakening considerably the flux actually passing through the secondary coil.

This condition of things may easily be explained by hydraulic analogy. Let, in Fig. 2, a ring-shaped tube of porous material be filled with and immersed in water, and let the water in the tube, as shown by the arrow, be kept in motion by a propelling fan I. This fan produces a difference of pressure between its inlet and outlet side, which pressure is absorbed by the frictional resistance of the tube. Since the pressure above the fan is greater than that below, water will, as indicated by the dotted lines, pass out through the pores of the tube in its upper half, and enter the tube through the pores in its lower half.

The velocity of the water must consequently be greater at *a* than at *b*. If the tube is wide and the propelling power of the fan small, little head will suffice to overcome the friction; and the quantity of water leaking out and in, as well as the difference of velocity in *a* and *b*, will be small. Let now a second fan (II) be inserted at *b*, which for the present we will imagine to be frictionless; then this fan will be set in rotation by the stream of water, but it will not increase the

Fig. 2.

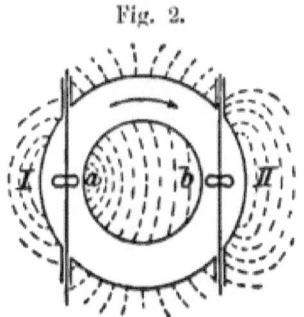

leakage nor diminish the velocity of the water. If, however, we retard the motion of fan II by letting its spindle transmit mechanical energy, the free flow of the water will be impeded, and the difference of pressure between the upper and lower halves of the tube will be increased. As a result, the leakage will be augmented, and the quantity of water passing the point *a* in unit time will be appreciably more than that passing the point *b*. At the same time the speed of fan II will be reduced; and this for two reasons. In the first place because the load on the spindle of II must retard its motion, and in the second place because the velocity of the water is smaller than before. If we wish to limit the loss of speed due to the latter cause, we can do so by placing the fan I as near as possible to fan II.

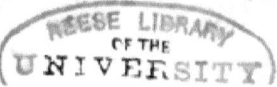

Now let us substitute for the porous tube the iron ring, and for the two fans the driving and the driven coil; then we see that the magnetic flux through the driven coil (which corresponds to the velocity of the water at h) will be the smaller the stronger the current is in the driven coil.

The arrangement of coils shown in Fig. 1 is bad, on account of their great distance. It does not give a strong magnetic flux through the driven coil if a large current is permitted to flow through this coil. We can improve the design by spreading the coils each over half the circumference of the ring, as shown in Fig. 3. In this case the

Fig. 3.

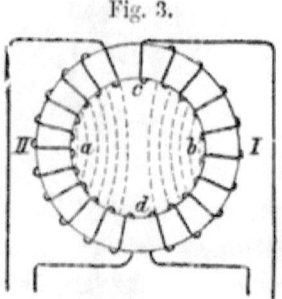

magnetic pressure tending to force the lines through the air is no longer constant over each half of the ring, but it attains its previous value only in the points c and d. It diminishes on either side of the vertical diameter, and becomes zero in a and b. The leakage field is therefore not only quantitatively smaller, but, owing to its distribution and the distribution of the two windings, its qualitative influence is also lessened, as compared with the arrangement shown in Fig. 1.

The distribution of the leakage field may be approximately determined if we remember that the magnetic pressure, which forces the lines to leave the ring at any point, is proportional to the ampère turns counted up to that point. Imagine now the windings evenly distributed, and the direction of the currents thus, that the magnetic pressure is from iron to air in the upper left quadrant, and from air to iron in the lower left quadrant. Corresponding pressures must of course exist in the right quadrants. Let now the ring be cut at *a* and straightened out, then the zig-zag line in Fig. 4 gives a graphic representation

Fig. 4.

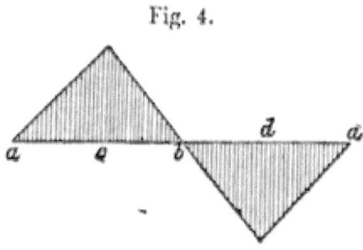

of the magnetic pressure producing leakage. Positive ordinates represent a pressure from iron to air, *i. e.* north polarity, and negative ordinates the opposite pressure, or south polarity. The leakage lines are shown dotted in Fig. 3, but only inside the ring. There are, of course, also leakage lines in the whole of the air space surrounding the ring. If we assume, as a very rough approximation, that the magnetic resistance along any path through air is the same, then the number of lines passing through unit surface in any point of the ring will be proportional to the magnetic pressure at that point, and the shaded areas in Fig. 4 may be taken to represent roughly the leakage field. The assumption that all the paths through air have the same magnetic resistance is of course not strictly correct; as we are, however, at present only concerned with a general investigation of the leakage field, it is not necessary to enter minutely into the question of how the magnetic resistance of any particular path through the air varies, and we may as a rough approximation assume that Fig. 4 represents the leakage field.

We have up to the present assumed that the two coils carry continuous currents, but it is obvious that our reasoning applies equally to the case of alternating currents, provided that the change in direction occurs in both coils nearly simultaneously, a condition which is always fulfilled in transformers when working under a load.

Arrangement of coils.—We have seen that in point of

Fig. 5.

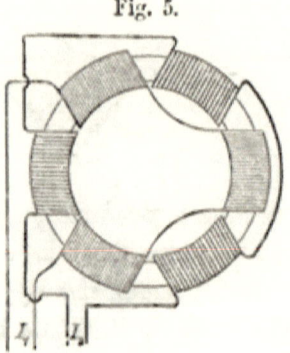

leakage Fig. 3 is an improvement on Fig. 1. We may, however, carry the improvement still further by sub-dividing each unit into several parts. In Fig. 5 we have six separate coils, arranged to uniformly cover the ring, and connected alternately with the primary and secondary circuit. The greatest magnetic pressure is also in this case at the junction of two coils; since, however, the number of turns in each coil is reduced to one-third, this pressure is also reduced to one-third of its previous value. The surface through which lines can leak is at the same time also reduced to one-third, so that the total leakage field now only amounts to $\frac{1}{3} \times \frac{1}{3} = \frac{1}{9}$ of its previous value. If, instead of subdividing each coil into three parts, we subdivide it into four, the leakage field would be reduced to $\frac{1}{16}$ of its previous value, and so on. It will thus be seen that, by carrying the principle of subdivision sufficiently far, we can reduce the leakage field to any desired extent. It could even be reduced to zero if we were to interweave the primary and secondary coils. This would, however, lead to difficulties as regards insulation, and is not necessary, since experience has shown that it suffices for all practical purposes to subdivide the windings so far as to limit the effective ampère turns in each individual coil to 500 or 600.

We have hitherto assumed that the magnetic link between the two windings is a circular ring, but it will

Fig. 6.

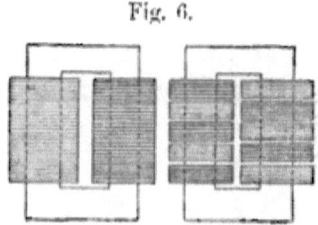

be obvious that its geometric form is immaterial, and that any shape of ring may be used. We might, for instance, employ a magnetic link in the shape of a rectangular frame, and place the coils over the two longer sides of the rectangle (Fig. 6). The arrangement shown on the left corresponds to Fig. 3. In this case there is only one secondary and one primary coil, and the magnetic leakage must therefore be very great. In the arrangement shown on the right, the primary winding is subdivided into five coils, which alternate in position with five coils of the secondary winding. The leakage is thereby reduced to about $\frac{1}{25}$ part. Another, and with regard to the reduction of leakage equally effective, arrangement consists in placing the coils axially within each other. This arrangement has the advantage of a reduction in the number of separate coils to be wound and handled, whilst at the same time the insulation between primary and secondary coils is of simple shape (plain cylinders), and can therefore easily be made perfect. ⌄

Fundamental equation.—The E.M.F. induced in a coil is, by a well-known law of electro-dynamics, proportional to the number of turns of wire, and the time-rate of change of the magnetic flux. In symbols—

$$E = n \frac{d N}{d t}.$$

In order to be able to calculate the E.M.F. occurring at any given moment of time, we must know the relation between N and t. The magnetic flux N is produced by the current passing through the primary coil, and if the magnetization of the iron core remains within such limits that the permeability may be considered to remain constant, then N may be considered to be proportional to the primary current. We assume for the present that the secondary coil is open, so that no current can flow through it which would mask or disturb the magnetizing effect of the primary current. There are, as a matter of fact, certain secondary actions which interfere with the strict proportionality of primary current and magnetic flux, but the consideration of these we must postpone. We also assume that the primary current is obtained from an alternator, the E.M.F. of which follows a true sine law. This is not always, and indeed very seldom, the case ; but it will be shown later on that the equations obtained under these assumptions remain applicable in all cases occurring in practice.

Imagine, then, a wire coil of one turn, including an area of *s.* square centimetres, traversed at right angles to its plane by a magnetic flux, which varies according to a periodic sine function between the limits $+ N$ and $- N$. The maximum value of the induction is obviously $B = N : s.$ Call the time required for the performance of a complete

Fig. 7.

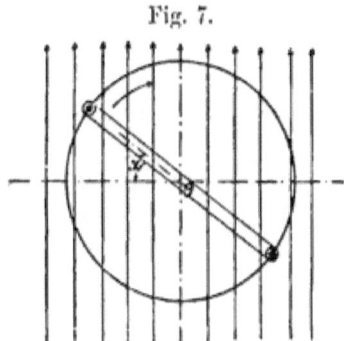

cycle from $+ N$ to $- N$, and back to $+ N$, T, and the number of cycles occurring per second $\sim$; then

$$T = \frac{1}{\sim}.$$

Since the E.M.F. is dependent on the change in the flux passing through the wire loop, but not on the angle at which the lines of force thread through the loop, we may replace the rectilinear and oscillating field by a constant and homogeneous field, provided we revolve the loop round an axis in its plane with a speed of $\sim$ revolutions per second. Let, in Fig. 7, the field be represented by the vertical lines, and O be the axis around which the wire loop is rotated. If the rotation takes place in the direc-

tion shown by the arrow, and if we count the time from
the moment the loop is horizontal, then let, at time t, the
coil occupy the angular position a. Through the cutting
of the lines of force there will be induced in the upper
half of the loop an E.M.F. directed towards the observer,
and in the lower half from the observer. In this and all
the following diagrams we mark these directions respectively by a dot and a cross inscribed into the little circle
representing the cross-section of the wire; these signs
meaning the point and the feathers of an arrow which
indicates the direction of E.M.F. or current.

Let ω be the angular velocity of the loop, then $a = \omega t$,
and $\omega = 2\pi \sim$; from which it follows that

$$a = 2\pi \sim t.$$

The magnetic flux threading through the loop is obviously $N \cos a$, and its rate of change

$$-\frac{d\, N \cos a}{d\, t} = N \sin a \, \frac{d\, a}{d\, t}.$$

Since $\frac{d\, a}{d\, t} = \omega = 2\pi \sim$, we have for the instantaneous

value of the E.M.F. the expression

$$E = 2\pi \sim N \sin a.$$

The loop has only one turn of wire. If there are n turns
the same E.M.F. is induced in each, and the total E.M.F.,
measured at the terminals of a coil of n turns, is therefore
in absolute measurement—

$$E = 2\pi \sim N n \sin a.$$

To obtain it in volts we must multiply by 10^{-8}. If the
coil is horizontal, the flux threading through it is a maximum, and the E.M.F. is zero ($a = r$). If the coil is

vertical, *i. e.* parallel to the direction of the field, the flux passing through the coil is zero, and the E.M.F. has its maximum value—

$$E = 2\pi \sim N n \, 10^{-8} \quad . \quad . \quad . \quad . \quad (1)$$

The instantaneous value of the E.M.F. is therefore

$$E_t = E \sin a,$$

and the instantaneous value of the magnetic flux is

$$N_t = N \cos a;$$

if by N we denote its maximum value.

These equations show that between E.M.F. and magnetic flux there is a difference in phase of a quarter period.

Imagine now the terminals of the coil connected with an incandescent lamp of resistance R. The current I_t passing through the lamp must vary as the E.M.F., E_t. Calling I the maximum value of the current, we should have $I = \dfrac{E}{R}$, and $I_t = I \sin a$. Although ordinarily the resistance of a lamp depends on and varies with the current, we are justified in assuming the resistance in our case to be constant, since the changes in current strength occur so rapidly that the filament has no time to grow hotter or cooler as the current grows stronger or weaker, but assumes a mean temperature, and has consequently a constant resistance. Let, then, the lamp be fed, first by our alternating current derived from the coil, and secondly by a continuous current, but let in both cases the filament be raised to the same temperature, so as to get the same amount of light. The lamp will then in both cases require the same supply of electric power, and we may consider the strength of the continuous current as a measure of the effective strength of the alternating current. Since $I_t = I \sin a$, we may represent the instantaneous value of an alternating current by the projection of a vector of length I, which revolves with an angular speed of $a = 2\pi \sim$. In the same manner other quantities of a periodic character may be represented, and the diagrams used for this purpose are called clock diagrams. Let then, in our case, the maximum value of the current be graphically represented to a certain scale by the length of a line I. Let the line revolve round one of its ends, and take the projection of the other end at stated times. The length of the projection, measured with the same scale, gives the instantaneous value of the current.

To find the total work which was supplied to the lamp by the current during the time of one period, we should divide the circle described by the current vector into a sufficiently large number of equal parts, distant by the time interval Δt from each other, project these points to get I_t, and form the expression $\Sigma I_t^2 R \Delta t$. This would be a laborious process, but it can be simplified if we imagine the additions made twice over by counting together such positions of the vector I as are 90° apart. The members of the series we have to sum up would then be of the form

$$R \, (I^2 \, sin^2 a + I^2 \, cos^2 a) \, \Delta t = R \, I^2 \, \Delta t,$$

as will easily be seen by reference to Fig. 8, in which the vector is shown in two positions differing by 90°. The projections are $O I_1$ and $O I_2$, and as the sum of their squares is obviously equal I^2, we find that each member of our series has the same value, namely $R \, I^2 \, \Delta t$. Let m be the number of members, so that $m = \dfrac{T}{\Delta t}$; then we find the total work done by the current during one period, by multiplying the value of one member of the series by m and dividing by 2, the latter because we have by taking the vectors in pairs counted them twice over. The work done during one period is

$$A = \frac{R I^2 T}{2}$$

and the power; that is, the rate at which work is done is

$$P = \frac{R I^2}{2}.$$

If the lamp is fed by a continuous current i the power is $P = i^2 R$. Let the power be the same in both cases, then

Fig. 8.

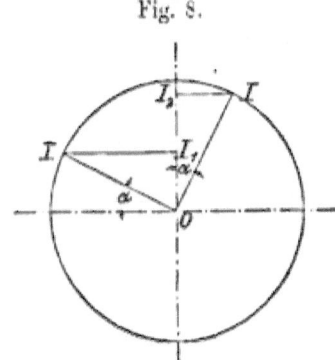

i may be considered the effective value of the alternating current, and will be given by the expression

$$i = \frac{I}{\sqrt{2}} \quad \ldots \ldots \quad (2)$$

This relation is of course only valid if the current is a sine function of the time. If it follows any other law the ratio between its maximum and effective value will be given, not by $\sqrt{2}$, but by some other co-efficient. If by

I we denote the instantaneous, and by i the effective, value of the current, we have for any form of current curve

$$A = \int_0^T I^2 R\,dt = T R i^2$$

$$i = \sqrt{\frac{1}{T}\int_0^T I^2\,dt} \quad . \quad . \quad . \quad . \quad (3)$$

In words: The effective current is the square root of the mean squares of the instantaneous values. $\checkmark$

The same reasoning applies, of course, to the pressure at the terminals of our coil, and in fact to any alternating E.M.F. Since in all instruments intended for the measurement of alternating pressure (hot wire, electrostatic and electro-dynamic) the action is dependent on the square of the E.M.F. applied to the instrument, the quantity actually indicated is the square root of the mean squares; or in symbols—

$$e = \sqrt{\frac{1}{T}\int_0^T E^2\,dt} \quad . \quad . \quad . \quad . \quad (4)$$

If E changes according to a sine law so that the instantaneous value $E_t = E \sin 2\pi \sim t$, where E is the maximum value of the E.M.F., then is the effective value

$$e = \frac{E}{\sqrt{2}} \quad . \quad . \quad . \quad . \quad . \quad (5)$$

It has been previously shown that $E = 2\pi \sim N n\, 10^{-8}$ represents the maximum value of the E.M.F. in volts induced in a coil of n turns through which the magnetic flux N passes, if the frequency is $\sim$ complete cycles per second. Combining equation 5 with this expression, we obtain

$$e = \frac{2\pi}{\sqrt{2}} \sim N n\, 10^{-8}$$

$$e = 4{\cdot}44 \sim N n\, 10^{-8} \quad . \quad . \quad . \quad . \quad (6)$$

This is the fundamental equation for the determination of the effective E.M.F. in the coils of a transformer, provided the E.M.F. curve is sinusoidal. For other shapes of the E.M.F. curve the equation will be given in the next chapter. Equation 6 applies, of course, to both the primary and secondary coils. If we distinguish the corresponding quantities by the indices 1 and 2, and assume for the present that the flux is the same in both coils, we have

$$e_1 = 4{\cdot}44 \sim N n_1\, 10^{-8}$$

$$e_2 = 4{\cdot}44 \sim N n_2\, 10^{-8}$$

These equations are therefore only applicable if there be no magnetic leakage. If there be leakage, a correction must be applied, as will be shown later on.

Since the current does work which is absorbed by the primary coil, the direction of e_1 must, on the whole, be opposed to the direction of the current. The secondary coil gives work to its external circuit, and the secondary current must therefore, on the whole, be of the same direction as e_2. The power supplied to or given off by the transformer can, however, not generally be considered to be correctly represented by the product effective current × effective pressure, since the phases of these two quantities are, as a rule, not coincident; that is to say, the current attains its maximum value at a different time from the E.M.F., and the times at which they pass through zero are also different. In the primary coil the product of instantaneous current and instantaneous E.M.F. is therefore not always negative, and in the secondary coil this product is not always positive; the work impressed on or given off by the transformer during the time of a period is therefore smaller than $T e_1 i_1$ and $T e_2 i_2$ respectively. The determination of the true work and true power will be given in Chapter IV.

CHAPTER II

Losses in transformers.—The losses occurring in transformers are of various kinds. There are first of all losses due to the ohmic resistance of the coil, causing so-called "current heat." These may be easily determined by Ohm's law, and need no further consideration here. Then there may be losses caused by eddy currents in the conductors, or other metallic parts of the apparatus. To calculate these is exceedingly difficult, and often impossible; on the other hand, it is always possible, by suitably placing or subdividing the metallic parts of a transformer, to reduce the eddy current losses so far as to make them a negligible quantity. Finally, we have to consider the losses occurring in the iron core, which are due to two causes: hysteresis and eddy currents.

If the induction in the core passes through a complete cycle from $+ B$ through zero to $- B$, and back through zero to $+ B$, a certain amount of electrical work is transformed into heat. This amount depends on the quantity and quality of the iron affected, and the maximum value of the induction. The corresponding power is proportional to the frequency, but is independent of the shape of the wave representing the induction, provided there is in each cycle only one maximum and one minimum. According to Steinmetz, the work lost per unit weight of iron per cycle is represented by an expression of the form

$$A = h B^{16},$$

where h is a co-efficient depending on the quality of the iron, and on the magnitude of the unit of weight we have chosen.

The magnetic flux surging to and fro within the mass of the iron produces in it E.M. forces, which in turn give rise to eddy currents, and thus cause loss. Let us assume an iron core of rectangular cross-section with sides a and δ; and let us compare different sections all of the same width a, but of different thickness δ, this dimension being, however, always small in comparison with a. The E.M.F. producing eddy currents must obviously be a maximum in the contour of the rectangle, that is, the skin of the core, and must be proportional to the whole of the flux passing through the cross-section; namely $B\,a\,\delta$. For a given value of B the E.M.F. varies therefore as $a\,\delta$, and the same holds for all the smaller E.M. forces which occur in the lower layers throughout the mass of the metal. The currents are inversely proportional to the resistances of the different layers, that is to say, the larger δ the smaller are all the resistances, and the stronger are all the eddy currents. With increasing thickness δ we have therefore a proportional increase in the E.M.F., producing eddy currents, and an increase in quadratic ratio of these currents themselves. The power lost increases therefore as δ^3.

For a core of circular cross-section the E.M.F. in each layer must obviously be proportional to the square of its diameter, whilst for layers of similar proportions (radial thickness of the layer a definite fraction of its diameter) the resistance is constant and independent of the diameter. The currents are therefore proportional to the square of the diameter, and the losses increase as the fourth power of the diameter.

In order to reduce these losses, it is therefore only necessary to reduce the diameter or the thickness δ, that is to say, the core must not be solid, but made up of wires or plates. In a wire core the loss is proportional to the fourth power of the diameter of the wire, and in a plate core to the third power of the thickness of the plates. By reducing the thickness to one-half or a third, the loss may thus be brought down to $\frac{1}{8}$ or $\frac{1}{27}$ of its previous value respectively. It is thus possible to make the loss negligibly small by using plates thin enough, but in practice the subdivision is not carried to the extreme limit, because the expense involved and the loss of space through insulation would outweigh the possible gain. It is sufficient to carry the subdivision to such a point that the eddy current losses become reasonably small, and this point has been found in practice to lie between a thickness of plates of 0·35 to 0·5 mm. (14 to 20 mils.). The stouter plates are used for frequencies of about 50, and the thinner plates for higher frequencies up to about 100. For a very low frequency and a low induction, plates thicker than $\frac{1}{2}$ mm. may be used. How far it is permissible to increase the thickness of plates can best be shown by an example. Let us assume that we have found in practice plates of ·5 mm. suitable for $\sim$ = 50 and B = 4,000, and that we have to design a transformer for $\sim$ = 20 and B = 3,000. How thick may the plates be made in order to have the same eddy current loss per

kilogram of iron ? In the transformer for 50 frequency the
E.M.F. which produces eddy currents is proportional to
$B \sim = 200,000$. If at the lower frequency and lower
induction we use the same thickness of plates, the E.M.F.
would be proportional to $20 \times 3000 = 60,000$, or only
·3 of the former value. Since the loss is proportional to
the square of the E.M.F., it would amount to only ·09 of
its former value. As, however, the same loss is permissible in the second case, we may increase the thickness
of the plates in the ratio

$$\sqrt[3]{\frac{1}{\cdot 09}} = 2\cdot 22.$$

We may thus use plates of 1·1 mm. (43 mils.) thickness,
and yet only lose as much power per kilogram of iron as
in the former case.

If, as is always the case in practice, the thickness of the
plates is such as to make the eddy current loss unimportant, then its exact and separate determination becomes
unnecessary. It forms only a small fraction of the
hysteresis loss, and may therefore be determined jointly
with the latter. It suffices for practical purposes to
measure the joint losses in sample plates of about the
usual thickness, and with inductions and periodicities
such as are generally employed ; the results may then be
used to predetermine the joint loss for thicknesses and
periodicities, not differing too widely from the respective
values obtaining in the experiments. I have made such
experiments with finished transformers and with samples
of plates, using for the latter purpose the special measuring
instrument which is described in Chapter VII., and have
thus determined the relation which exists between the
loss of work per cycle at different inductions. The thick-

ness of the plates varied between ·4 and ·5 mm. (16 and
20 mils.), and the periodicity between 50 and 75. Since
eddy current losses form only a very small fraction of the
hysteresis losses, the variation of the periodicity between
the narrow limits of 50 and 75 does not materially
influence the magnitude of the joint loss. The results of
my measurements are graphically represented in Fig. 9.
For convenience this figure gives, not the loss of work
per cycle, but the loss of power calculated for a frequency
of 100 cycles per second. For a different frequency the
losses must be proportionately altered. Two curves are

Fig. 9.

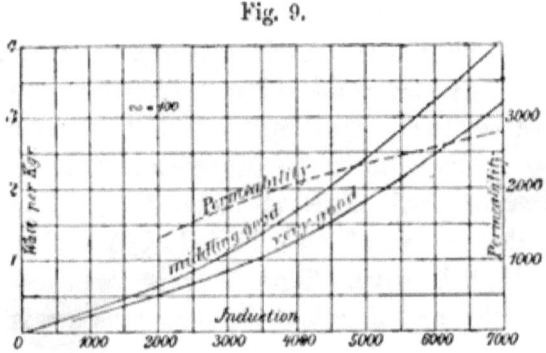

given; the upper represents ordinary wrought-iron plates
of fair quality, the lower plates of extra good quality,
specially rolled for use in transformers. The dotted
curve gives the permeability, which has been found by
measuring the primary current and loss of energy in
transformers when working an open circuit, as will be
more fully explained later on. The permeability has
been found for both qualities of iron to be approximately
the same.

Influence of the E.M.F. curve on hysteresis loss.—The

shape of the E.M.F. curve has not only an influence on
the relation between the effective and maximum E.M.F.,
but also on the hysteresis loss which necessarily ac-
companies the production of any given effective E.M.F.
If we assume that the flux has only one maximum in
each half period, then the hysteresis loss is dependent on
this maximum and the frequency, but is independent of
the shape of the E.M.F. curve, and of the shape of the
curve which represents N as a function of the time. We
may imagine a series of N, t curves which all have the
same maximum, but are otherwise of different shape.
The hysteresis loss with all these will be the same, but
not the effective E.M.F. Of these curves we would
naturally prefer that which gives the greatest E.M.F. for
any given maximum value of the flux. Since the flux is
dependent on the shape of the E.M.F. curve, we may
state the problem also in the following terms. Let there
be at our disposal different alternators, all of which
produce the same effective E.M.F., but with different
forms of E.M.F. curve. That form is to be selected in
which the flux, and therefore the hysteresis loss is least.

To solve this problem it is of course necessary to
assume certain forms of E.M.F. curve. As a starting-
point we may conveniently select the sine curve, and then
determine what influence an alteration in its shape has
on the relation between maximum and effective E.M.F.,
and on the hysteresis loss. The shape may be altered in
two ways; we may either flatten the curve, or make it
steeper. If we carry the flattening process to its theoretically (though not practically) possible limit, we obtain
a broken line consisting of vertical and horizontal parts.
The vertical parts represent the instantaneous change
from $- E$ to $+ E$, and the length of each horizontal

part represents the time of half a period. Such a curve might be obtained by the commutation of a continuous E.M.F., but only as long as no appreciable current is allowed to flow. With an alternator of special design, the rectangular shape of E.M.F. curve might also approximately be obtained, though never completely. We are therefore justified in regarding this shape as the extreme limit of the flattening out of the sine curve with which we started. The effective E.M.F., e, is then equal to the maximum E.M.F., E, or in symbols—

$$e = E.$$

Since E is a constant for each half period, $\dfrac{d N}{d t}$ must during that time be also a constant, and the flux must be represented by a zig-zag line, with joints having the same abscissæ as the vertical parts of the E.M.F. line. From Fig. 10 it will be seen that

$$\frac{d N}{d t} = \frac{4 N}{T}$$

Since for one turn $E = \dfrac{d N}{d t}$, and since $e = E$, we have for n turns

$$e = 4 \sim n N\ 10^{-8} \quad . \quad . \quad . \quad . \quad . \quad . \quad (7)$$

For a sine curve we have by (6)

$$e = 4.44 \sim n N\ 10^{-8} \quad . \quad . \quad . \quad . \quad . \quad (6)$$

If, then, the effective E.M.F., e, is to be the same in both cases, then the induction N must, with a rectangular E.M.F. curve, be greater than with a sine curve in the ratio of $4.44 : 4$; that is to say, the maximum value of the flux N must be by 11 per cent. greater, or with the same induction B we must put 11 per cent. more iron into the core. The hysteresis loss is therefore increased by at least 11 per cent. As already pointed out, the rectangular form is an extreme case, scarcely attainable in practice. In reality, the curve will assume a shape roughly represented by the dotted line where the sharp corners are rounded, and thus the peaks of the N, t curve will also be rounded and lowered. The hysteresis loss will therefore be somewhat smaller than corresponds to the extreme case, but certainly greater than with a sine curve. Our investigation thus far has therefore shown

Fig. 10.

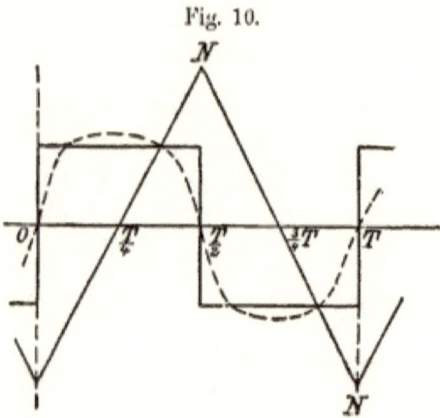

that an alternator giving a flat E.M.F. curve is disadvantageous for working transformers.

Let us now see what the conditions of working are with a peaky E.M.F. curve. In this case the limit is not definable. We may imagine the E.M.F. curve composed of a series of inclined straight lines, thus forming a succession of triangles; but we may also imagine this form exaggerated, i. e. the lines convex towards each other, like the sides of a tent, whereby the peaks would be raised higher than with the triangle form. As a matter of fact,

certain alternators, in which the armature is provided
with projecting teeth, give an E.M.F. curve having very
steep and high peaks, so that E is very large in comparison with c. We shall not attempt to investigate a
curve of this kind, since the investigation is extremely
complicated, and for our present purpose not required.
All we care to learn is whether a more pointed curve
than the sine curve is better or worse as regards hysteresis
loss, and for this purpose we may take the simple case of
a triangular shape. If we find that this gives us a
smaller hysteresis loss than a sine curve, we may conclude
that the exaggerated triangle must, in this respect at
least, be still better.

Let, then, line E in Fig. 11 represent the E.M.F. given
by the alternator; what will be the curve of induction?
Since for one turn

$$E_t = \frac{d N}{d t}$$

in absolute units, the curve in question must be such that
its trigonometric tangent in any point a, whose abscissa
is t, is equal to the ordinate of the E.M.F. line having the
same abscissa. In symbols—

$$E_t = t \, \tan a = - \frac{d N_t}{d t};$$

$$N_t = - \int t \, \tan a \, d t + \text{constant};$$

$$N_t = \text{constant} - \frac{1}{2} t^2 \tan a.$$

The constant can be found from the condition that for
$t = 0, N_t = N$. We thus obtain

$$N_t = N - \frac{1}{2} t^2 \tan a,$$

the equation of a parabola. Since $E = \frac{T}{4} \tan a$, we have also

$$N_t = N - \frac{1}{2}\, t^2\, \frac{4\,E}{T}.$$

For $t = \frac{T}{4}$, $N_t = 0$; from which we find

$$N = \frac{1}{2}\left(\frac{T}{4}\right)^2 \frac{4\,E}{T}$$

$$N = \frac{T\,E}{8}\ \text{in absolute measure.}$$

Since $\frac{1}{T} = \sim$, we have

$$E = 8 \sim N\, 10^{-8}\ \text{volts.}$$

We see thus that with a triangular E.M.F. line the

Fig. 11.

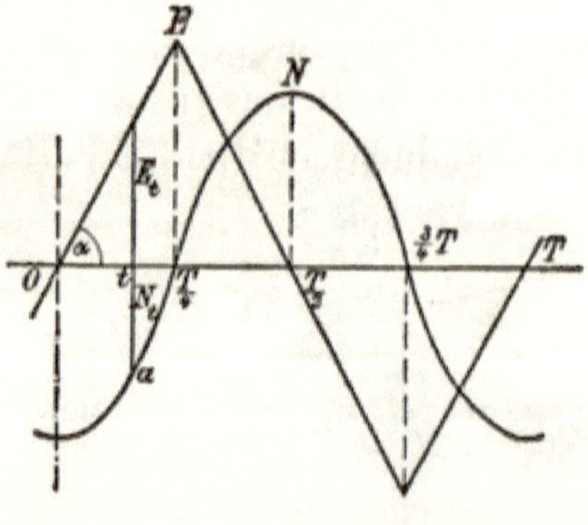

maximum E.M.F. is, for the same induction, exactly doubled as compared with the rectangular form, and the stress on the insulation is therefore also doubled. The point of interest is, however, not the maximum, but the effective value of the E.M.F.

This is for a quarter period given by the following equation—

$$c = \sqrt{\frac{4}{T} \int_0^{\frac{T}{4}} E_t^2\, d\,t}$$

$$E = t \tan a$$

$$c = \sqrt{\frac{4 \tan^2 a}{T} \int_0^{\frac{T}{4}} t^2\, d\,t} = \sqrt{\frac{4 \tan^2 a}{T} \frac{1}{3} \frac{T^3}{64}}$$

$$c = \frac{T}{4} \tan a\ \frac{1}{\sqrt{3}}$$

$$c = \frac{E}{\sqrt{3}} \ i\ E = c \sqrt{3}.$$

By inserting this value in the above equation for E we find

$$c = 4\cdot 62 \sim N\ 10^{-8} \text{ for one turn, and}$$
$$c = 4\cdot 62 \sim n\ N\ 10^{-8} \quad . \quad . \quad . \quad . \quad . \quad . \quad (8)$$

for a coil of n turns.

Since the co-efficient is greater than with a sine curve, it follows that a smaller induction N with the same quantity of iron, or a smaller quantity of iron with the same induction, will now suffice to produce the desired effective E.M.F.

The equations (6), (7), (8) have all the form

$$c = c \sim n\ N\ 10^{-8},$$

and differ only in the co-efficient c, which has the following values—

(1) With a rectangular E.M.F. curve . . $c = 4\cdot 00$
(2) „ „ sine „ „ . . . $c = 4\cdot 44$
(3) „ „ triangular „ „ . . . $c = 4\cdot 62$

Imagine now that we have three alternators with E.M.F. curves of these shapes, and let us supply the same

transformer successively with currents from these machines,
and at the same E.M.F. The maximum induction N will
be different in each case. It is greatest in case 1, and
smallest in case 3. If now we take the iron loss in case
2, where the E.M.F. is represented by a sine curve as our
standard of comparison, and call it unity, then the loss
will be—

With a rectangular E.M.F. curve . . . . . . 1·11
With a triangular E.M.F. curve . . . . . 0·96

The last figure shows that there is, as regards hysteresis
loss, some, though not an overwhelming advantage in em-
ploying a machine giving a peaky E.M.F. curve; whilst, on
the other hand, machines with a flat E.M.F. curve are
disadvantageous.

Influence of the shape of core and coils upon the losses.—
Since the hysteresis loss is proportional to the weight of
iron, we must aim at making this as small as possible. In
designing the iron core of a transformer, we are limited by
two conditions. First, the iron of the core must suffice to
pass the total flux N with a moderate induction; and
secondly, its length must suffice for housing the coils. On
the other hand, it is desirable to make the length of each
turn of wire as short as possible, in order to reduce the
current heat. These conditions are in part contradictory,
and cannot each and all be fully met. The best design is
consequently merely a compromise, and can only be
obtained by a method of trial and error carried to a point
at which any further change in dimensions or winding does
not reduce the sum of all the losses.

The shape of the cross-section of the core is of great
importance for the length of wire required in the coils and
the resistance. Thus, a rectangular shape is worse than a

square, since it requires more length of wire for the enclosure of an equal area. For the same reason, a circle is better than a square of equal area, though if for constructive reasons the diameter of the circle cannot be made greater than the side of the square (*i.e.* the condition of equal area cannot be fulfilled), then a slight advantage lies with the square core, which contains $\frac{4}{\pi}$ times the amount of iron as compared with a circular core. Let r be the radius of the circle, and $2r$ the side of the square, δ the thickness of the insulating covering on the core, and d the depth of the winding. For the same induction B and the same number of turns the E.M. forces will be in the ratio of $\pi r^2 : 4 r^2$ for the circular and square core respectively. The mean length of winding is, for the circular core, $\pi (2 (r+\delta)+d)$, and for the square core $8 (r+\delta)+\pi d$. The E.M.F. induced per unit length of winding is therefore proportional to $\pi r^2 : \pi (2 (r+\delta)+d)$, and $4 r^2 : (8 (r+\delta)+\pi d)$, and the ratio between these values is

$$\frac{2 (r + \delta) + d}{2 (r + \delta) + \frac{\pi}{4} d} > 1 ;$$

from which it will be seen that with a square core the E.M.F. induced per unit length of winding is greater than with a round core, and that the difference becomes the more marked the greater the depth of winding. The explanation for this paradoxical result (that a square core requires less wire than a circular) lies in this, that the two cores have not the same area. The square core contains more iron, and the electrical output is greater. The larger apparatus has, of course, an advantage over the

smaller which more than compensates for the less advantageous shape of the core.

It has already been mentioned that the core of a transformer must be built up of wire or plates, in order that
eddy currents may be avoided. If wire be used for this
purpose, no special insulation is required, and the space
actually filled by iron is from 78 to 80 per cent. of the
total. When plates are used, they must be insulated
against each other, though no specially good insulating
medium is required. The insulation may consist of a
layer of oxide on the plate itself, or a coat of varnish, or
paper insertion. The latter method of insulating is the
most reliable, and occasions a loss of space of from 12 to 15
per cent., so that about 87½ per cent. of the total space is
actually occupied by iron. The available space is, therefore,
better utilized with plates than with wire, and for this
reason, as well as on account of the more mechanical construction which is possible with plates, the latter are
nearly always used in the building up of transformer cores.

Core and shell transformers.—As already explained, the
action of a transformer is due to the linking together of
two coils by means of an iron core. This principle of inter-
linking can be carried out in a variety of ways; one of the
simplest is represented in Fig. 6. The link is a rectangular
frame, and the coils are placed upon the two longer limbs
of the rectangle. Such an arrangement is called a *core
transformer*, and is characterized by the fact that most of
the iron is within the coils, whilst the external surface of
the coils is everywhere exposed to the cooling action of
the air.

We may, however, also change the relative position of
iron and copper. We may assume that the rectangular
frame is composed of copper wire, forming the two coils,

which are laid close upon each other, and the external cylindrical parts of Fig. 6 of iron discs, or a winding of iron wire which forms a kind of iron shell in which parts of the coils are imbedded. Such a transformer is called a *shell transformer*.

Whether the core or the shell type is better cannot be generally decided, but depends on a variety of circumstances. In the core type, the weight of iron is small and the turns of wire short. On the other hand, the number of turns is great (because N is small), and thus the total weight of copper is, notwithstanding the small perimeter (length of one turn), fairly large. The length of the path of the flux is also great, and the ampère turns required to produce the flux are large. On the other hand, we have the advantage that the coils are accessible, and exposed to the cooling effect of the air.

The shell type has the advantage of a short magnetic path, which requires but few ampère turns to produce the flux; the coils have fewer turns, and the total weight of copper is, notwithstanding the larger perimeter, small. It has, however, the disadvantage of requiring considerably more iron, whilst the coils are not accessible, nor so well exposed to the cooling effect of the air.

In order to compare in a general way the two types of transformer, we may show what effect alterations in the arrangement of iron and copper have in a particular case. For this purpose, we may start with any concrete design and alter it in various ways, but always under the condition that the induction in the iron, current density in the copper, and total output shall be the same. For the sake of simplicity we may assume that the same gauge of wire shall be used in all cases, so that the number of turns in each coil will be proportional to the winding space. The current

D

and pressure being the same for all designs, it follows that the flux must be inversely proportional to the winding space; and as the induction is to be the same, it further follows that the area of the core and that of the winding

Fig. 12.

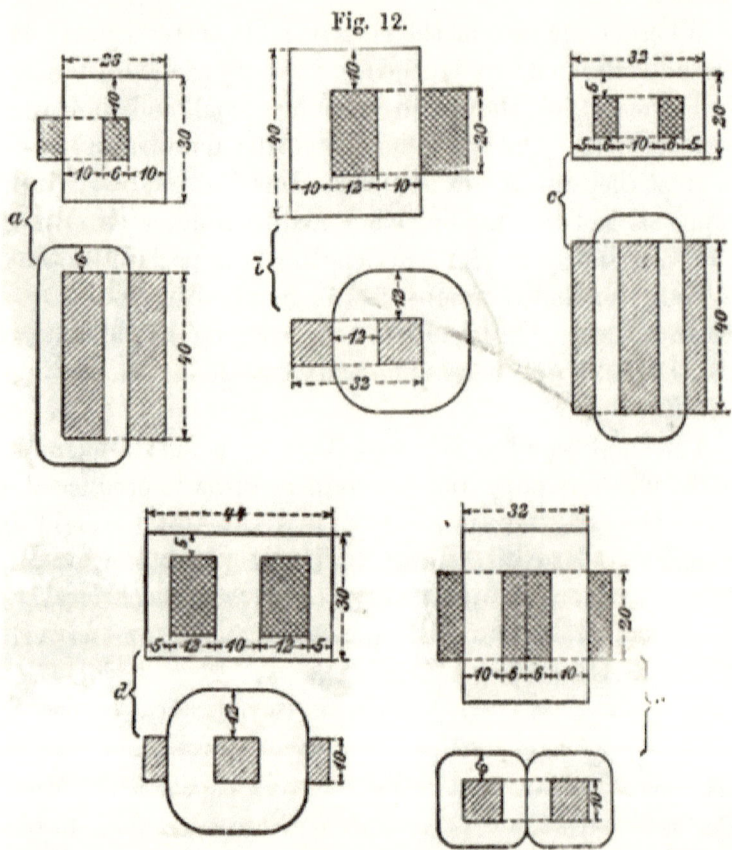

space are also inversely proportional. Under these conditions we may roughly judge the merit of each design by the weight of iron and the length of wire.

As a starting-point, we may take the design shown in Fig. 12*a*, in which the area of the core is 400 sq. c.m.

(inclusive of the space wasted, in insulation), and that of the winding space 60 sq. c.m. The iron weighs 200 kgr., and the mean perimeter of the winding is 119 c.m. With 100 turns of wire we have thus a total length of 119 m. Let us now alter this design by reducing the core area to one-quarter of the previous value and quadruple the winding space. The iron part will then be 40 c.m., but only 10 c.m. long, in a direction at right angles to the plates, and the winding space must now be 12 × 20 c.m. We arrive thus at the design Fig. 12*b*. The mean perimeter of the coils is now only 78 c.m., but as we must have four times as many turns as before, the total length of wire has been increased to 312 m., that is nearly three times the former length. On the other hand, the quantity of iron has been much reduced; it is only 73 kgr. If the iron is cheap but of good quality and copper is dear, then type *a* is preferable. With cheap copper and bad iron type *b* will be the better design.

Both of these designs may, however, be still improved. We may, for instance, so alter type *a* as to embed both sides of the coil in iron and thus arrive at a true shell transformer, Fig. 12*c*. This does not alter the length of winding, which is still 119 m., but it reduces the weight of iron considerably. The flux through the coil divides now to both sides, and only half the previous area is required in the outside shell. The iron weight is thus reduced to 112 kgr. We thus obtain the design of shell transformer now largely used in England and America.

By making similar alterations in Fig. 12*b* we arrive at the type Fig. 12*d*, which is also a shell transformer, but lacks the advantage of a short length of winding. This length is as before 312 m.; on the other hand, the weight of iron has been reduced to 59 kgr. This design is only justifiable if no iron of good quality can be obtained, and there is no need to be economical in copper. In former times, before rolling mills were capable of turning out such first-rate transformer plates as now, there was some reason for building transformers of this type. Now-a-days, however, excellent iron can be obtained from a variety of mills, and there is no reason to practise economy in iron at the cost of an increased weight of copper. It is therefore preferable to alter the design Fig. 12*b* in such way as to save copper, and this can be done by dividing the winding and placing it upon both limbs. We arrive thus at the type, Fig. 12*e*, which is a true core transformer. The perimeter of the winding is on account of its smaller depth considerably reduced, and the iron weight is not excessive. It is 73 kgr., and the length of wire is 236 m. This type of transformer is much used in England and Germany.

For convenience of comparison the above results are summarized in the following table :—

Type.			Weight of Iron. Kgr.			Length of wire. Meters.
a	...	...	210	...	...	119
b	...	...	73	...	...	312
c	...	...	112	...	...	119
d	...	...	59	...	...	312
e	...	...	73	...	...	236

In all these types the magnetic path lies completely in iron. There exists, however, also another type of transformer in which the lines of force pass only partly through iron and close themselves through air. This is the so-called *hedgehog* transformer, Fig. 13*a*, introduced by Swinburne with the intention of reducing the iron loss.

For this purpose Swinburne winds the coils upon a core consisting of a bundle of iron wires with their ends spread out like the back of a hedgehog. The lines of force pass then from one end of the core to the other, through air, as shown by the dotted lines, and hysteresis loss only takes place in the small quantity of iron which forms the core proper; in the shell of air surrounding the coils there is, of course, no hysteresis loss. This type of transformer has not been successful in practice. If we imagine two such transformers placed side by side and the ends of the iron wires bent so as to meet (Fig. 13*b*), we obtain an ordinary core transformer, the hysteresis loss in which can

Fig. 13.

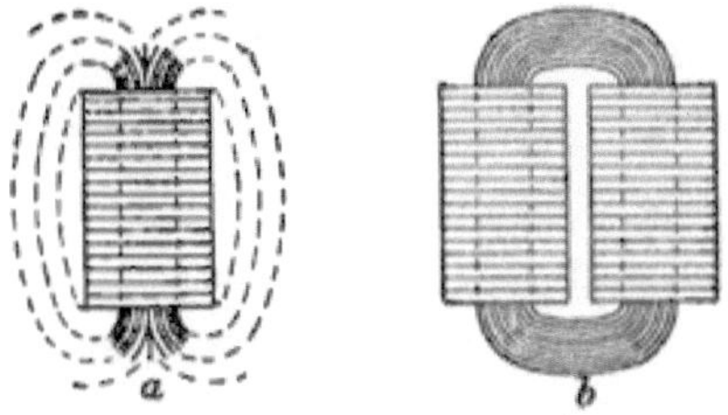

only be very slightly greater than in two single hedgehogs, the small increase being due to the extra length of iron wire required to make a perfect junction between the two cores. On the other hand, there must be an increase in the hysteresis loss, in each core of the two single hedgehogs, because the induction in the middle of the core is considerably greater than at the ends. The E.M.F. is of course proportional to the mean induction and the hysteresis loss to the $\sqrt[1.6]{}$ of the mean values of $B^{1.6}$. It is therefore clear that any variation in the value of the induction along the core must result in a greater hysteresis loss than in the case where the induction is constant

throughout the length of the core. The hedgehog transformer has also another drawback, namely, that it requires
an exceedingly large primary current at no load. Whilst
in the types shown in Fig. 12 the no load current is but
a small fraction (a few per cent.) of the full load current,
the hedgehog takes with an open secondary up to 60 per
cent. of the full load primary current. This is a property
which renders the hedgehog transformer unfit for use in
central station work where a small day current is of the
utmost importance. There is however one purpose for
which this type of transformer is very suitable, namely, as
a choking coil, whereby its capability of letting large
currents pass under moderate E.M.F. is the very thing
desired. For transformer work proper the types shown in
Fig. 12 are however far preferable, especially the designs
12c and 12e.

CHAPTER III

Usual types.—The designs commonly used belong all to the two great groups of shell and core transformers. The former are of the kind shown in Fig. 14, where P and S

Fig. 14.

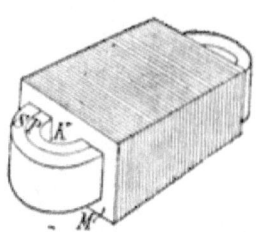

are the primary and secondary coils somewhat oblong in shape and placed either within or upon each other, whilst the iron part consists of rectangular plates each with two openings in which the winding is embedded so that only the rounded ends of the coils remain accessible. In a variety of this design the coils are circular, and the iron part is arranged in sections placed all round the circle. Beyond a slight increase in the cooling surface, this form has no advantage over the more usual form shown in Fig.

14. The coils are first wound and the iron plates are afterwards threaded over them in a manner to be presently described. The centre part of the plates K forms the core proper which carries the flux through the coils, whilst the external part M forms the shell.

Fig. 15.

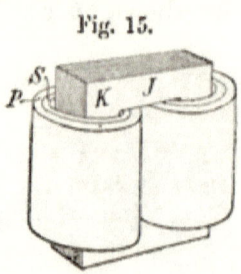

In the core transformer the iron part has the shape of a rectangular frame (Fig. 15), the longer limbs forming the cores K, and the shorter limbs the yokes J. The primary and secondary coils (P and S) are cylindrical, and may be placed within each other as shown in the figure, or as a

Fig. 16.   Fig. 17.

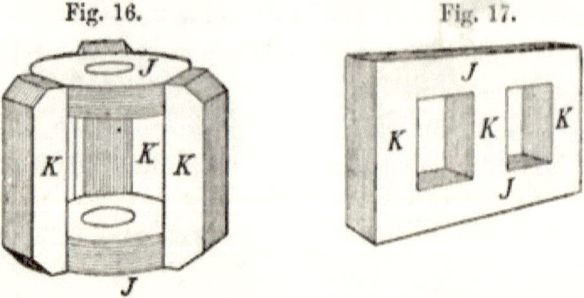

pile of narrow discs upon each other. For polyphase work the core type of transformer is almost exclusively used, and Figs. 16 and 17 show two arrangements of the iron part for this purpose. In the former the yokes J are formed of circular iron discs (armature discs may conveniently be used), and the three cores K are placed at an angular distance of $120°$ from each other and pressed against the yoke discs by top and bottom plates of cast-iron with slanting brackets engaging the tapered ends of the core pieces. The coils are placed over the cores as in Fig. 15. In Fig. 17 the true symmetrical arrangement is abandoned; the three cores K being in line and joined by top and bottom yokes J precisely as in Fig. 15. Each core is surrounded by the primary and secondary coil belonging to one phase. The advantage of this design lies in the fact that the yoke and core plates are parallel and not at right angles to each other, which renders the construction mechanically easier and magnetically more perfect.

Construction of the iron part.—In Figs. 14, 15 and 17 the plates forming the iron part are shown as complete surfaces without any joint between that part which forms the core and that which forms the yoke or shell. The use of such plates is possible, and would have the advantage that no interruption of any kind is offered to the flow of the lines. On the other hand, there would be the great disadvantage that the coils could only be wound after the iron part is built up. The wire would have to be threaded through the openings, and could therefore not be wound on a lathe. Moreover if the wire is stout it could scarcely be wound properly by hand, and to obtain a proper and reliable insulation would be next to impossible, whilst a fault in the insulation could only be detected after the transformer is completely finished. These are such serious drawbacks that the use of closed plates is not commercially possible. It is better to arrange the iron part in such way that the plates may be inserted after the coils are wound and tested, and for this purpose some sort of joint in each plate is necessary. It is true that the joint interrupts the continuity of the magnetic path and is in this respect imperfect, but the imperfection can with a suitable arrangement of plates be so far reduced as to be without appreciable influence on the magnetic conductivity of the iron part taken as a whole. For this purpose the plates are so arranged and laid upon each other that the joint in one plate is covered by the solid part of the next. The lines of force instead of leaping across the joint can then in part pass through the solid metal of the neighbouring plates, and since the surface through which they pass from

Fig. 18.   Fig. 19.

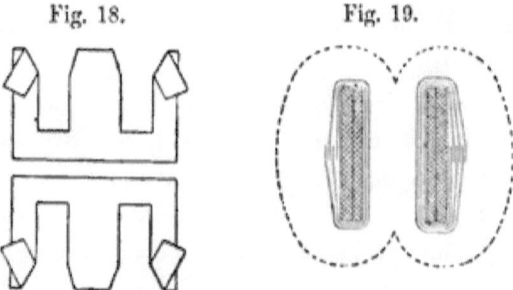

one plate to the other is immensely greater than the sectional area of each plate, the magnetic resistance of this bye-pass becomes negligible, and we may regard the iron part so built up as practically equivalent to one having no joints whatever.

The principle here explained may be illustrated by the way the iron parts of some of the usual types of transformer are built up. In the *Westinghouse* transformer the plates are stampings of the form shown in Fig. 18. Inclined slits are made to both sides of the central bridge piece forming the core, and the two parts of the shell are bent up as shown. In this condition the plate is inserted into the coils and the bent-up pieces are bent back so as to lie flat; the next plate is then inserted from the opposite side and its bent-up pieces laid flat. In this manner the joints in each plate are covered by the solid parts of its two neighbours, and the continuity of the magnetic path is thereby secured.

The core in *Ferranti's* transformer (Fig. 19) consists of a bundle of straight iron plates which is inserted into the coils. The plates are then one by one doubled back to form the shell. The length of the plates is such that

Fig. 20.

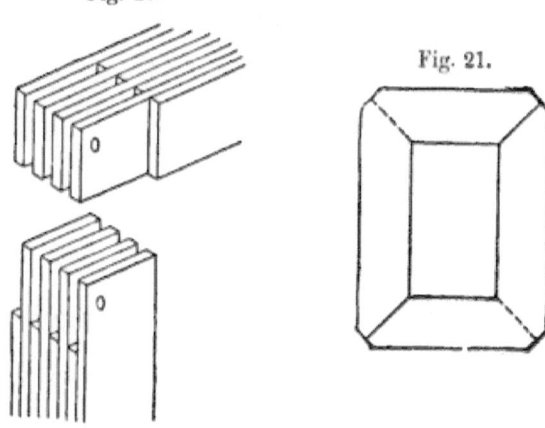

Fig. 21.

at the joint they overlap slightly, and thus form an easy path for the magnetic flux.

In the *Author's* transformer (Fig. 15) the plates for the core and the yoke are straight, and are dove-tailed into each other at their junction as shown in Fig. 20. In order to show the construction more clearly the thickness of the plates is much exaggerated in this figure. After the iron part is built up it is held together at the corners by insulated bolts. Since all the plates are rectangular there is no waste of material in cutting out.

In *Crompton's* transformer (Fig. 21) the plates are cut

out in ∟ form and inserted into the coils alternately from one side and the other so as to cover the joints in one layer by the solid parts of the next. In cutting or punching the plates there is on account of the special shape some loss of material.

In stamping the plates for the Westinghouse transformer there is also some loss of material, namely, that punched out to form the two windows. To obviate this loss *Mordey* has altered the design in such way as to utilize the material punched out as the bridge-piece or core, Fig. 22. Each punching yields two pieces, the shell which is a rectangular frame and the core which is laid across it. The shell

Fig. 22.   Fig. 23.

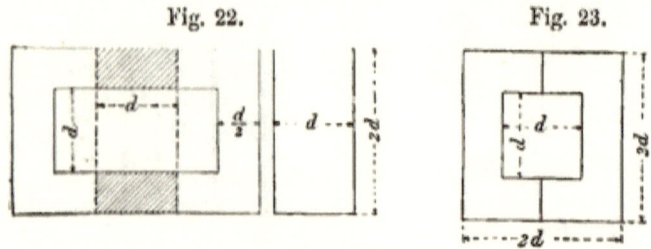

plates are placed over and the core plates through the coils. With the plates made in this way the winding space is of course dependent on the thickness d of the core, and cannot be chosen arbitrarily as in Fig. 18. The height of each window is d and its width $\frac{d}{2}$, as will be seen by Fig. 22. The external dimensions of the shell are $3d$ and $2d$. Contact between core and shell plates only takes place over the shaded areas; in all other parts there is a clearance between neighbouring plates equal to their thickness. In consequence there is only half the space within the coils actually filled by iron, and to carry the same flux the perimeter of the coils must be greater than in Fig. 18, where the whole space is filled by iron. To remedy this defect, Mordey uses additional stampings of the form shown in Fig. 23, which can also be done without any waste of material. The internal square of side d is used to fill up the interstices in the core of Fig. 22, and the two side pieces to fill up the interstices in the shell. There are thus five pieces required for every two layers, and all the material cut up is also utilized. Although the proportions resulting from this method of building up the iron part are suitable, it is sometimes desirable to vary slightly the dimensions of the winding space. This may be done whilst still retaining the Mordey system of punching, but a slight waste of material is then unavoidable.

In all the methods of building up described above, the principle of breaking joint is maintained and the iron part is magnetically equivalent with one having no joints. It is however also feasible to abandon this principle and build up the iron part with joints between core and yoke or core and shell. In Fig. 16 such joints are unavoidable; in transformers of the type shown in Figs. 15 and 17 they are sometimes used to facilitate the fitting up. It is then possible to completely finish each part of the apparatus and to piece it together with very little labour. In case of repair a coil may be taken out and replaced without having to take the plates out one by one as is the case with transformers constructed as described above. There are thus from a practical point of view many advantages in the use of joints; there is however the disadvantage that the no load—or idle—current becomes greater, as will be shown in the next chapter, and that unless the joints are very carefully made there is loss of power through eddy currents at the joint. To make the magnetic path as perfect as possible, it is of course desirable to reduce the

distance between the edges of the plates at the joint as far
as practicable. It is however not practicable to reduce it
to zero, partly for obvious mechanical reasons, and partly
because direct contact between the edges would give rise
to eddy currents. With the design shown in Fig. 16 this
is at once obvious. The plates cross at right angles, and if
there were contact between them there would be formed
closed circuits through which considerable eddy currents
would flow and produce heat whilst wasting power. To
avoid this defect a sheet of insulating material must be
inserted at the joint, and the latter can therefore not be
magnetically perfect.

Even where the plates do not cross, but are parallel to

Fig. 24.

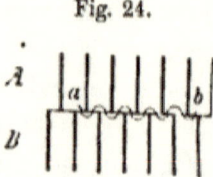

each other, the insertion of a sheet of insulating material
is desirable, as will be seen on inspection of Fig. 24, which
represents a joint, metal to metal. The thickness of the
plates is shown exaggerated, and the stout black lines
between the plates represent the insulation. Although
the thickness of the plates is supposed to be the same, we
cannot expect that at the joint insulation and metal will
register perfectly. There must be portions on the surface
where there is a slight displacement, so that the insulation
in part *A* touches not the insulation but the plate in part
B, as shown in the figure. It is obvious that in this case
there is metallic connection between plates *a* and *b*,
and that eddy currents as indicated by the wavy line must

flow. To interrupt the path of such currents we must separate parts A and B by inserting a sheet of insulation as shown in Fig. 25. Thus the magnetic resistance of the joint must be increased in order to avoid heating and loss of power.

Proportions of the iron part.—The value of a transformer design is largely dependent on the proportions of the iron part. If, for instance, the cores in Fig. 15 are made very short, the yoke pieces must be made correspondingly long in order to obtain sufficient winding space. At the same time, the depth of winding, and consequently the weight of copper is increased, and we thus make the

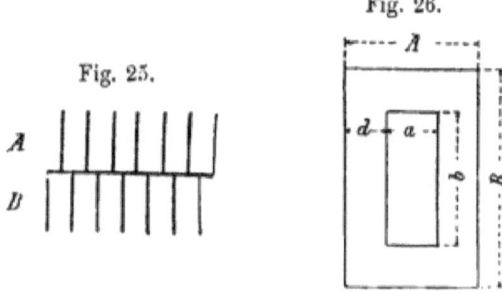

Fig. 25.

Fig. 26.

design worse by shortening the cores. On the other hand, if to get a small perimeter for the winding we make the cores very long and thin, we get a long magnetic path, and consequently a large no load current. We shall do better to increase the core section, and reduce the number of turns, but not too much, because of hysteresis loss. No hard and fast rules can be given for the best proportion, and the only way to determine them is by a method of trial and error in a series of tentative designs, paying due regard to such questions as the desired efficiency, the average and maximum loads, the cost of iron and copper, etc. As a starting-point for the design, the proportions shown in Fig. 26 may be taken. The section of the cores and yokes is supposed to be a square, and if the junctions are made by dove-tailing, the edges of the cores may be chamfered so as to reduce the mean perimeter of the winding. It is convenient to bring all dimensions into relation with the thickness d of the core given in millimeters. We thus have as a first approximation to a good design—

$$a = 10 \; + 1{\cdot}2d$$
$$b = 100 + 2{\cdot}6d$$
$$A = 10 \; + 3{\cdot}2d$$
$$B = 100 + 4{\cdot}6d$$

Fig. 27.

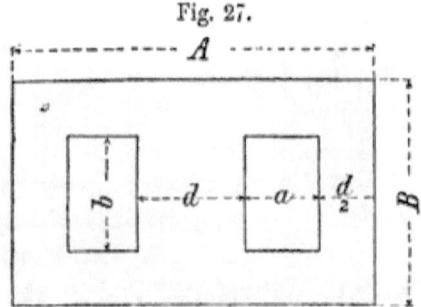

For shell transformers with short cores, the plates of which are to be punched without any loss of material, we have (Fig. 27)—

$$a = \frac{d}{2} \qquad b = d$$
$$A = 3d \qquad B = 2d$$

If however a slight waste of material is to be permitted, the dimension a may be somewhat increased so as to

get more room for winding and generally better proportions.

$$a = 0{\cdot}6d \qquad\qquad b = d$$
$$A = 3{\cdot}2d \qquad\qquad B = 2{\cdot}2d$$

or

$$a = 0{\cdot}7d \qquad\qquad b = d$$
$$A = 3{\cdot}4d \qquad\qquad B = 2{\cdot}4d$$

The depth of the core and shell measured at right angles
to the plane of the plates may be $2d$ to $4d$.

Heating of Transformers.—If a transformer is at work,
a part of the energy supplied to it is transformed into
heat which must be dissipated by radiation or convection.
A necessary condition for this dissipation is an increase
of temperature of the transformer over its surrounding
medium, and this increase is naturally the greater the
smaller the surface as compared with the power lost.
The rise in temperature is thus a function of the $\sigma = \dfrac{S}{P_v}$
if by S we denote the total external surface of the transformer, and by P_v the power transformed into heat, that is
lost between the primary and secondary. The character
of this function can only be determined experimentally,
and varies of course with the type of the transformer
tested. Is the arrangement such that the external air
has free access to the iron part and the coils, the cooling
effect is greater than in transformers placed in a case.
The rise of temperature $T = f(\sigma)$ will therefore be
smaller in the former and larger in the latter case. If
the transformer is placed in a case the rise of temperature
will be greater if the case contains air only, than if it is
filled with some insulating liquid, such as oil, which conveys the heat more easily to the walls of the casing.

E

The use of oil as a filling-in material, provided it is absolutely free from moisture, has also the advantage of maintaining the insulation when the transformer is erected in the open air or in damp places. Since the load is subject to fluctuations, the temperature of the transformer as well as that of the air in its case must vary. Thus the air in the case expands and contracts slowly, and in this way damp air from the outside is drawn in and may in time injure the insulation. This danger is avoided if the case be filled with oil, but since the temperature co-efficient of oil is great, provision must be made for its expansion, either by not filling the case completely, or providing a stand-pipe.

The heating of transformers is due to three causes: hysteresis, current heat, and eddy currents. The hysteresis loss is dependent on the mass of iron, the induction and the frequency, but is independent of the load. The heat produced by hysteresis is therefore constant if pressure and frequency are constant, which is nearly always the case in practice. This heat is developed in the iron. The current heat is developed in the copper, and is of course proportional to the square of the load. The heat produced by eddy currents may be produced not only in the iron, but also in the coils and in other metallic parts, especially if the winding is so arranged as to produce considerable magnetic leakage. It is however always possible, and for other reasons generally necessary, to so arrange the winding as to reduce magnetic leakage as much as possible. The eddy currents are then unimportant, and the loss occasioned by them becomes negligible.

We need therefore only consider the heat produced by hysteresis and ohmic losses. The former may be determined by measuring the power supplied to the primary circuit when the secondary circuit is open; the latter may be calculated for any load by Ohm's law if the resistances of the two circuits are known. We thus find the total loss. To determine the corresponding rise of temperature we must run the transformer under the corresponding load and measure its temperature at intervals. It is advisable to use a spirit—and not a quicksilver—thermometer for this purpose, since if there should be a little magnetic leakage the metal in the bulb of the thermometer would itself become the seat of eddy currents, and get heated on

Fig. 28.

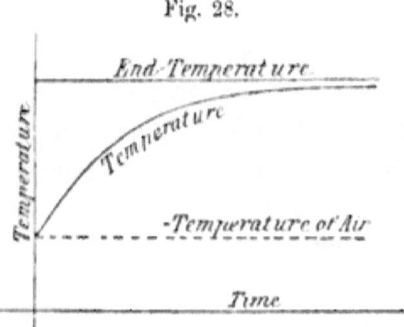

that account. If then we plot temperature as a function of time, we obtain a curve of the general character shown by Fig. 28. The curve rises rapidly at first, and then approaches a horizontal asymptote, the ordinate of which is the final temperature. This final temperature is reached slowly; with small transformers only after some hours, and with large transformers of high efficiency it may take several days. To carry out the experiment in this way would not only require much time but the expenditure of considerable power over a long period, and be therefore costly. To avoid this difficulty we may use the expedient of working the transformer light, but supplying to it that amount of power which it would lose at full load. This may be done in one of two ways: we may work it light at higher voltage so as to expend the whole of the power in hysteresis, or we may send a continuous current through one of the windings and regulate its pressure so as to expend the whole of the power in current heat. In the first case we heat only the iron and in the second only a part of the copper, so that neither method of heating corresponds exactly to the conditions of regular work. By combining both methods we obtain however a sufficiently close approach to these conditions. It is convenient to do the heating by alternating current during the day when the machinery of the works is in motion, and that by continuous current during the night when a storage battery may be used.

Results of tests.—I have in this manner made an extended series of tests with various sizes of transformers and at various loads to find the relation between rise of temperature and cooling surface per watt lost. All the transformers tested were cased in, and the cases sometimes filled with oil. The results are given in the two curves (Fig. 29). The transformers stood on a cemented floor in a large covered space, so that the air had access from all sides, and some of the heat could also flow away through the floor. Had the transformers been tested in the open, the rise of temperature would have been somewhat smaller; if they had been tested in a confined space it would have been somewhat greater than shown by these curves. Where the case was filled with oil no mechanical means were used to circulate the oil.

In using the curves (Fig. 29) it should be remembered that they refer to transformers worked an indefinitely long time at full load. Most transformers, and especially those for lighting purposes, are continuously at work, but not always under full load. This circumstance should be considered in the determination of the cooling surface, which must be provided so that the temperature of the transformer shall not exceed a certain limit. The losses

Fig. 29.

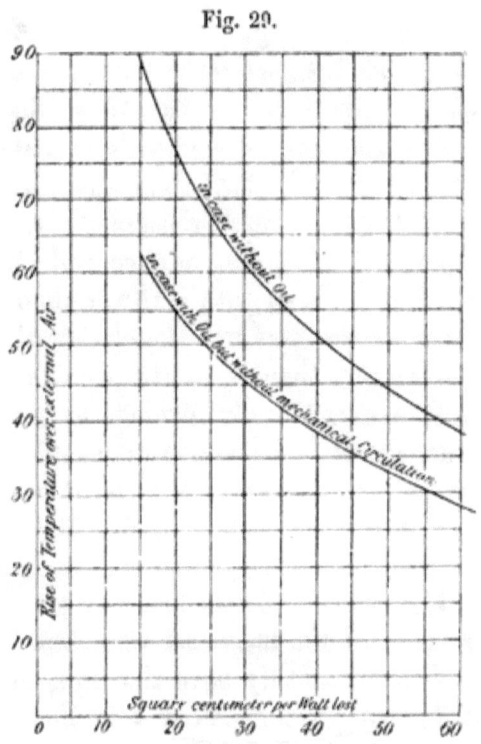

at various periods during the twenty-four hours can be separately determined, and thus an average for the loss P_v can be found which inserted into the formula $\sigma = \dfrac{S}{P_v}$ P_v, gives the correct abscissa to be used in Fig. 29.

Influence of the linear dimensions.—Machinery and

mechanical apparatus generally become, as a rule, more efficient when made on a larger scale, and judging by this fact we should expect large transformers to have an advantage over small, both as regards the efficiency as also the amount of material required per Kilowatt output. Within certain limits this is indeed the case. A 10 Kwt. transformer weighs less than 10 transformers of 1 Kwt. each, and its efficiency is also higher. But if we compare a 100 Kwt. and a 10 Kwt. transformer, the advantage of the larger apparatus especially as regards weight is not so great, and may under certain conditions vanish altogether, the reason being that the larger apparatus has relatively to its weight or bulk a smaller cooling surface. If the power lost per unit of cooling surface and therefore the temperature is to be kept within the original limits, the magnetic stress in the iron and the electric stress in the copper must be less than with the smaller apparatus; hence the advantage generally incident to mere size may in part or entirely be lost because of insufficient cooling surface. As long as the smaller of the two transformers is of such a size that the cooling surface is largely in excess of what is required, then the large transformer may be worked at a higher and yet safe temperature and full advantage be taken of the increased size; but if the small transformer is itself already near the limit of output as determined by temperature, then this limit must also determine the output of the large transformer, and the better utilization as regards the material is marred by the less favourable conditions as regards cooling surface.

To illustrate by an example. Let us assume we have a 10 Kwt. transformer which shows at full load a loss of 400 watts; namely 200 watts hysteresis and 100 watt current heat in each circuit. The total cooling surface is 16,000 sq. c.m. or 40 sq. c.m. per watt, giving a temperature rise of a little over 50° C. Let the induction be 5000. Now let us double the linear dimensions. With the same induction we would get four times the flux, and to get the same E.M.F. we should only require one-fourth the number of turns in each circuit. But the winding space is four times as great as before, so that each wire with its insulations may now have sixteen times the area. Since the insulation need not be increased in the same proportion as the wire, the latter may have a little more than sixteen times the area. At the same time we may use rectangular wire in the high-pressure coil (which with the smaller transformer would have been impracticable), and thus also gain somewhat in conductivity. These are advantages incidental to the increase in size of apparatus, but in order not to complicate our comparison we shall for the present disregard them, bearing however in mind that the results we obtain for the large transformer will be less favourable to it than is actually the case.

Let us first consider the low-pressure winding. The mean perimeter has been doubled, but the number of turns has been reduced to one-fourth. We have therefore half the length of wire. Had we used the same gauge of wire we would have half the resistance; since however the section of the wire is 16 times what it was before, the resistance is only $\frac{1}{32}$ of what it was before. Let the small transformer be wound for 100 volts secondary pressure. With 100 ampère at full load and a loss of 100 watts as assumed, we should have a resistance of ·01 ohm in the secondary circuit. The secondary of the large transformer would have a resistance of $\frac{.01}{32}$ ohm, and at the same current density would carry 1600 ampère with a loss in pressure of ·5 volt. The same reasoning applies to the primary

winding, so that the total loss in current heat is 1 per cent.
of full load, or 1600 watts. The hysteresis loss has increased
with the weight of iron to eight times its former value,
and amounts also to 1600 watts. We have thus a total loss
of 3200 watts, or 2 per cent. of the output against 4 per
cent. in the smaller transformer. The large transformer is
therefore, as regards efficiency, very much better than the
small, provided we can really load it to 160 Kwt. This
however is only possible if we make special provision for
carrying away the heat. We have only four times the
cooling surface and eight times the loss of power. The
cooling surface per watt of loss is therefore not 40 sq. c.m.
as before, but only 20 sq. c.m., and the rise of temperature
in air without oil or other means of cooling would be 76°
C., which is inadmissible. With oil the temperature rise
would be 55° C., and this may be permitted.

To work the small transformer in air and the large in oil
is however not a fair comparison; to put both on the same
footing we must work the large transformer also in air, and
that is only possible if we so far lower the magnetic and
electric stress as to reduce the total loss to 1600 watts,
which will give a temperature rise of 50° C. The hysteresis
loss with an induction of 5000 was 1600. From the curve
Fig. 9 we find that for very good iron each kilogram at B
$= 5000$, and $\sim = 100$ absorbs 1·8 watt. In order to halve
the loss we must drop the induction to such a value that
the loss is only ·9 watt. From the curve we find that this
is the case for $B = 3150$. The flux and therefore the
E.M.F. is now reduced in the ratio of 5000 to 3150, that
is to say we shall only get 63 volts on the secondary and
not 100 as before, provided we do not alter the winding.
To reduce the current heat also to one-half we must reduce
the current in the ratio of $\sqrt{2}$ to 1. The secondary cur-

rent will therefore be 1600 : $\sqrt{2}$ = 1135 Ampère; and the output will be 63 × 1135 = 71·5 Kwt. This is the theoretical result, obtained whilst disregarding the advantage resulting from the fact that in the larger transformer the winding space can be better utilized. This advantage may be considerable, but can only be determined by making the design. In the present case we should gain about 8·5 Kwt., making 80 Kwt. as the output which the large transformer can give at the same temperature rise as the small transformer. The output is thus 8 times and the weight also 8 times as great as before, so that the utilization of the material is no better in the large than in the small apparatus. As regards efficiency, however, the increase in linear dimensions has produced a sensible improvement. The small transformer has a loss of 4 per cent. and an efficiency of 96 per cent. The large transformer has an output of 80 Kwt., and a total loss of 1600 watts, or 2 per cent. The efficiency is thus not 96, but 98 per cent.

The advantage resulting from increase in linear dimensions becomes more important if the output is not limited by the ratio of total cooling surface and loss of power. This is the case; either if the small transformer from which we start is so small that its output is only limited by the question of efficiency and not by heating; or if the large transformer is provided with special means for cooling it. If such means can be devised we obtain with the same magnetic and electric stresses an output proportional to the fourth power of the linear dimensions (in our example 10 × 2⁴ = 160 Kwt.), whilst the weight only increases as the third power. The utilization of the material is therefore the better the larger the transformer; in other words: by increasing the linear dimensions we decrease the weight of material per Kwt. output. The loss of power taken as a percentage of the output is at the same time inversely proportional to the linear dimensions.

We may even go a step further. If we succeed in devising some specially effective cooling arrangements (such as a mechanical circulation of the coil in combination with cooling the same by water in a separate vessel) the magnetic and electric stresses may in the large transformer be even greater than in the small transformer (which is devoid of such cooling arrangements), and the output may be increased in even a greater ratio than the fourth power of the linear dimensions. Let us, for instance, increase the pressure in the transformer previously considered by so much that $B = 7300$. The hysteresis loss is then 3·5 watts per kgr. at $\sim = 100$, or in all $\frac{3·5}{1·8}·1600 = 3100$ watts.

The E.M.F. of the secondary will be increased from 100 to $100 \times \frac{7310}{5000} = 146$ volts. If we allow an increase in current heat equal to that of hysteresis we may increase the current output in the ratio of $1 : \sqrt{\frac{3100}{1600}} = 1 : 1·39$. The output will therefore not be 160 Kwt. but

$$\frac{146 \times 1600 \times 1·39}{1000} = 325 \text{ Kwt.}$$

with a total loss of 6200 watts, or nearly 2 per cent. of the output. The efficiency is thus the same as before, whilst the output is doubled. The cooling surface is now however not 40 sq. c.m. but only 10 sq. c.m. per watt of lost power, and this transformer could only be safely worked if specially effective cooling arrangements were provided. With small transformers such arrangements would be too costly, both in the first outlay and in working, but with large transformers the cost becomes insignificant as compared with the saving of material, and it pays to provide special cooling appliances.

The above considerations were made on the supposition that the linear dimensions of the two transformers under comparison are in the ratio of 1 : 2. It would of course also have been possible to take any other ratio 1 : m, but the concrete example facilitated the calculation, and having by means of an example ascertained the relations, we may translate them into the general case of an increase of linear dimensions in the ratio of 1 : m. It must however be remembered that the results are only approximately correct ; and this for two reasons : First, because we have been obliged to take as a starting-point a definite induction in the small transformer; and secondly, because the skill of the designer is a factor of importance which cannot however be included in our calculation. The result is therefore only a rough approximation, but as such useful when getting out a preliminary design. For this purpose the results are tabulated below for the general case of the linear dimensions of a small standard transformer being increased in the ratio of 1 : m.

The output of the standard transformer is P, and its loss of power P_v; its cooling surface is S and the cooling surface per watt loss is $\sigma = S : P$. The same symbols with the index 1 refer to the large transformer.

We may conveniently distinguish three cases.

I. The cooling surface per watt loss is the same in both transformers, as are also the method of cooling and the temperature rise.

II. The magnetic and electric stresses are the same in both transformers. With the same method of cooling the large transformer must get hotter; or if the same temperature rise is required the large transformer must be cooled more efficiently.

III. The magnetic and electric stresses in the large transformer are so far increased that the cooling surface per watt lost is reduced in the ratio of $1 : m^2$. The cooling arrangements in the large transformer must then be especially efficient.

DESIGNATION OF QUANTITIES.	STANDARD TRANSFORMER.	LARGE TRANSFORMER.		
		I.	II.	III.
Linear Dimensions ...	l	$l_1 = ml$	ml	ml
Weight	G	$G_1 = m^3G$	m^3G	m^3G
Cooling Surface ...	S	$S_1 = m^2S$	m^2S	m^2S
Output	P	$P_1 = m^3P$	m^4P	m^5P
Loss	P_v	$P_{v1} = m^2P_v$	m^3P_v	m^4P_v
$S : P_v$	σ	$\sigma_1 = \sigma$	$\dfrac{\sigma}{m}$	$\dfrac{\sigma}{m^2}$
Weight per Kwt. output	g	$g_1 = g$	$\dfrac{g}{m}$	$\dfrac{g}{m^2}$
Efficiency	normal	larger	larger	larger

The use of this table can best be explained by means of an example. Let us assume that we have got out a good design for a 5 Kwt. transformer with an efficiency of 95 per cent. We consider this our standard transformer, and wish to get out an equally good transformer for 15 Kwt. on the same lines. To set out the drawing we must in the first place know by how much the linear dimensions will have to be increased. Both transformers shall be cooled by air only. In the standard transformer $\sigma = 60$, and the temperature rise is 38° C. In the large transformer we allow a temperature rise of 55° C., to which form the curve Fig. 29 corresponds $\sigma_1 = 35$. If we assume (to start with) equality of stresses (column II. of the table) we obtain m

$$= \frac{\sigma}{\sigma} = 1\cdot72$$ and a transformer of too large an output,

since $P_1 = 5 \times 1\cdot72^4 = 43\cdot5$ Kwt. We may now consider the transformer of 43·5 Kwt. to be the standard transformer, and find from column II. the dimensions of the 15 Kwt. transformer. We have now $P = 43·5$ and $P_1 = 15$. This gives

$$m = \sqrt[4]{\frac{15}{43·5}} = 0·77$$

The linear dimension of the 5 Kwt. transformer was l, that of the 43·5 Kwt. was $1·72\ l$. The linear dimension of the 15 Kwt. is therefore

$$l_1 = l \times 1·72 \times 0·77 = 1·325\ l$$

If thus the thickness of the core in the 5 Kwt. transformer is 100 m.m. we should set out the drawing for the 15 Kwt. transformer with a core 133 m.m. thick.

As a check on this calculation we may also determine the thickness of the core from column I. for equal heating, that is for $\sigma_1 = 60$. We have then $P_1 = m^3 \times 5 = 15$ and

$$m = \sqrt[3]{3} = 1·44$$

For equal heating the core would not be 133 but 144 m.m. thick, the difference arising from the fact that the temperature rise will now only be 38° C. and not 55° C. as permitted. With $\sigma_1 = 60$ the weight of the transformer is $G^1 = 3\ G$; with $\sigma = 35$ and a temperature rise of 55°, the weight would be $G_1 = 1·325^3\ G = 2·33\ G$. If we allow this temperature rise, we can therefore reduce the weight of the 15 Kwt. transformer and approximately also its cost by $\dfrac{300 - 233}{100} = 22·4$ per cent.

The cooling surface of the 5 Kwt. transformer was 60 sq. c.m. per watt loss, or in all $S = 250 \times 60 = 1500$ sq. c.m. For the large transformer with a core of 133 m.m. we have $S_1 = 15,000 \times 1·33^2 = 26·500$ sq. c. m. or 35 sq. c.m. per watt

loss. The total loss is $26,500 : 35 = 760$ watts, or about 5
per cent. of the output. If however we use the stouter
core of 144 m.m. we lose $P_{e1} = 1·44^2 \times 250 = 520$ watts,
or only 3·5 per cent. of the output. We may tabulate
the above results as follows, assuming as a basis for the
weights that the small transformer weighs 20 kgr. per
Kwt. output.

DESIGNATION OF QUANTITIES.		TRANSFORMER OF	
	5 Kwt.	15 Kwt.	15 Kwt.
Thickness of core m.m. ...	100	133	144
Weight kgr.	100	233	300
Weight per Kwt.	20	15·6	20
Temperature rise °C. ...	38	55	38

The designer in setting out the drawing has now to
decide whether to make the core 133 or 144 m.m. To
make it stouter than 144 m.m. would make the transformer
too heavy, whereas a thinner core than 133 m.m. would
necessitate the employment of a special cooling apparatus,
the cost of which would be excessive as compared with the
saving in the cost of the transformer itself. The choice
of dimensions lies therefore within the limits given above,
and must be made merely with reference to the relative
value of efficiency on the one hand and low cost on the
other. If a cheap and light transformer is required, the
designer will decide for the smaller dimensions; if high
efficiency is the first consideration, he will choose the
larger dimensions.

The practical value of the above considerations regarding
the influence of the linear dimensions on the output lies in
this, that the designer is thereby furnished with a method
to determine, without waste of time, the dimensions leading to a good design, if, to start with, he has the drawing
of a good standard transformer as a pattern to go by.

CHAPTER IV

POWER OF ALTERNATING CURRENT—COMBINING CURRENTS OR PRESSURES
—DETERMINATION OF NO LOAD CURRENT—INFLUENCE OF JOINTS.

Power of alternating current.—In order to investigate the working condition of a transformer we must be able to determine the power given to the primary and taken from the secondary terminals. It is therefore necessary that we should be able to find either by direct measurement, or in some other way, the power conveyed by an alternating current. We assume for the present that current and E.M.F. follow a sine curve. This assumption is made for the sake of simplicity. It is not always correct, but we shall see later on that the methods of measuring power which are based on this assumption are also applicable in the general case where the current as well as the E.M.F. follow any law, provided the frequency of both is the same.

Let, in Fig. 30, the sine line I represent the current as a function of the time, and the line E, the E.M.F. impressed on any two points of the circuit, say, for instance, the primary terminals of a transformer. We count the time in the direction to the right. At the time O the current is negative (the ordinate of the current curve I being below the axis), and the E.M.F. is zero. At time t_1 the current is zero and the E.M.F. has a certain positive value. The maximum E.M.F. occurs at time t_2 and the maximum current a little later at time t_3. At time t_4 the E.M.F. has decreased to zero, but the current is still positive, though rapidly decreasing. It reaches zero at time t_5, when the E.M.F. has already a negative value. Since both curves follow the same line the horizontal distances between their maximum and zero values must all be the same, that is to say, the time interval between any two

Fig. 30.

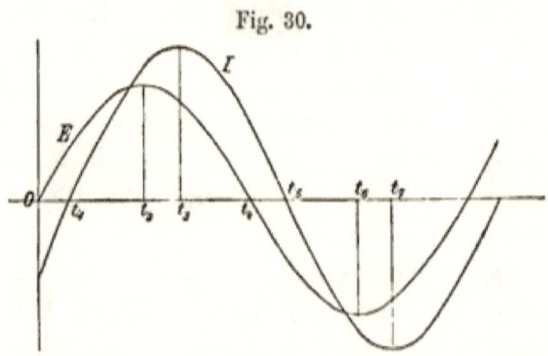

pairs of corresponding points is a contrast. Thus $t_3 - t_2$ $= t_5 - t_4 = t_7 - t_6$, etc. This time difference between corresponding values of the E.M.F. and current is called the lag or lead of current, or E.M.F. respectively. In our example where the E.M.F. passes its zero and maximum values before the current passes through the corresponding values, we have a lagging current as compared to the E.M.F.; or a leading E.M.F. as compared to the current. The condition under which this relation obtains is the existence of, in addition to the impressed E.M.F., of a second E.M.F. which tends to oppose any and every change of current. Such an E.M.F. is produced by the change in the magnetic flux due to the current. If, however, instead of this opposing E.M.F., there acts an E.M.F. in the inverse sense, then every change in current strength is thereby promoted, and the current attains its zero and maximum values sooner than the impressed E.M.F., or in other words. we have a current leading before the impressed E.M.F. Such a second E.M.F. tending to advance the current is produced by the insertion of a condenser into the circuit. The condenser takes the maximum positive charging current in the moment that the impressed E.M.F. passes through zero in a positive sense. When the impressed E.M.F. has attained its positive maximum the condenser is fully charged, and the charging current is zero. When the impressed E.M.F. now begins to decrease it is still positive, but the condenser begins already to discharge, producing a negative current, which becomes a maximum at the moment when the impressed E.M.F. passes through zero, and so on. We see thus that the condenser current leads over the impressed E.M.F. by a quarter period.

In addition to the two cases here considered, a third case is possible in which no second E.M.F. either advancing or retarding the current is acting; in this case (glow lamps fed from a transformer) the current will have the same phase as the impressed E.M.F., and its strength will be simply determined by Ohm's law.

The periodic variation in current and E.M.F. may be conveniently represented by a clock diagram. Let, in Fig. 31, the outermost circle be used to mark the time (somewhat in the fashion of a clock-dial), and let Ot be the hand of a clock revolving with constant angular speed. We count the time from the moment in which Ot stands horizontally to the left. Let in this moment the E.M.F. be zero. Describe a circle the radius of which represents

F

to any convenient scale the maximum value of the E.M.F., then the projections of this radius on the vertical gives to the same scale the instantaneous value of the E.M.F. at the time to which their position corresponds. Thus at the time t the E.M.F. vector occupies the position OE, and the instantaneous value at the E.M.F. is OE_t. We count the E.M.F. as positive if E_t is above, and negative if E_t is below the axis.

The instantaneous value of the current may be represented in a similar manner, but the current vector must

Fig. 31.

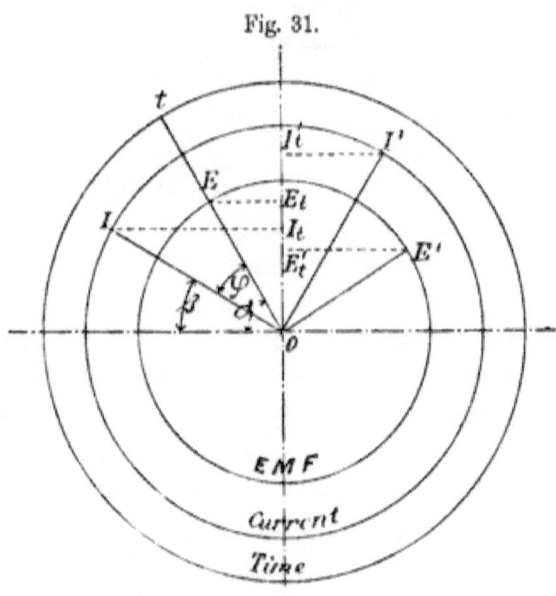

be drawn with an angular lag ϕ behind the vector of E.M.F. Both vectors revolve in the same sense, and with the same speed, thus always preserving their relative angular position. Let the vector OE occupy the position a at time t, then the current vector will occupy the position $a - \phi$, and the instantaneous values of E.M.F.

and current are $I \sin a$ and $E \sin (a-\phi)$ respectively. The instantaneous value of the power is—

$$P = EI \sin a \sin (a-\phi).$$

If the vectors perform $\sim$ revolutions per second so that $T = \dfrac{1}{\sim}$ is the periodic time, and if by ω we denote the angular speed, the following equations obtain—

$$\omega T = 2\pi$$
$$a = \omega t$$
$$d a = \omega d t$$
$$d a = 2\pi \sim d t$$

Since work is the product of power and time, we have

Fig. 32.

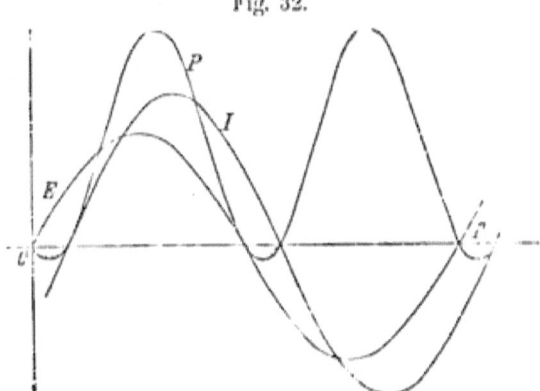

for the work performed by the current in the time dt the expression—

$$d A = P dt$$

In Fig. 32 are drawn the curves E and I to represent respectively E.M.F. and current. By multiplying their ordinates we obtain the ordinates of a third curve marked P, which represents the instantaneous value of the power whilst the area enclosed between P and the horizontal represents work. For ordinates above the horizontal, the power is positive, or given to the circuit; for those below the horizontal it is negative, or taken from the circuit. To obtain the work given to the circuit during a complete cycle, we must measure the area of P between the ordinates for $t = o$ and $t = T$, counting the small shaded parts below the horizontal as negative. The work corresponding to a complete cycle is—

$$A = \int_0^T P\, dt$$

The instantaneous power varies, as will be seen from Fig. 32, between a small negative and a larger positive maximum. Let us now suppose that we substitute for this varying power the constant power of a continuous current, so that the work taken over the time T is the same in both cases, then the constant power (which in future we call effective power) is the quotient of work and time, or in symbols—

$$P = \frac{A}{T}$$

$$P = \frac{1}{T} \int_0^T P\, dt$$

Substituting for P, dt, and T the values given above, we obtain also—

$$P = \frac{\sim}{2\pi \sim} \int_0^{2\pi} E\, I \sin a \sin (a - \phi)\, d a,$$

or after integration—

$$P = \frac{1}{2} E\, I \cos \phi$$

The same result can be obtained graphically, as was first shown by Blakesley, by using the diagram Fig. 31.

Let E and I represent the instantaneous positions of the vectors, then $E \sin a \; I \sin \beta$ represents the instantaneous value of the power. To find the effective power we would have to draw the vectors in a number of positions, determine the sum of all the expressions $E \sin a \; I \sin \beta$, and divide by the number of positions. Instead of performing this operation for neighbouring positions of the vectors in their right sequence, we may do so for a series of positions which are $90°$ apart. In this way we would count the power twice over, and to get the true value of the effective power we must divide the result by 2. The effective power will therefore be given by the expression—

$$P = \frac{1}{2\,m} \Sigma \left(E \; I \sin a \sin \beta + E \; I \cos a \cos \beta \right)$$

m being the number of positions counted in their natural sequence. The expression in brackets is $E \; I \cos (a - \beta)$, or $ET \cos \phi$, and is independent of the particular position or value of a chosen. All members of the series have therefore the same value, and the sum is simply $m \; E \; I \cos \phi$. We thus obtain

$$P = \frac{m \; E \; I \cos \phi}{2\,m}$$

$$P = \frac{1}{2} \; E \; I \cos \phi$$

the same expression as before.

E and I are the maxima of E.M.F. and current respectively. Between these and their effective values we have, as already shown, the relations $c = \dfrac{E}{\sqrt{2}}; \quad i = \dfrac{I}{\sqrt{2}}$

The effective power may therefore be expressed by the equation—

$$P = c \; i \cos \phi$$

In this expression e and i denote the effective values of E.M.F. and current respectively, and ϕ the lag between current and E.M.F. For a given E.M.F. and current we have maximum power, when $\phi = o$, that is when E.M.F. and current coincide in phase, whilst for $\phi = 90°$ the power is zero.

Since $i \cos \phi$ represents the projection of the current vector on the E.M.F. vector, we may represent the power graphically by the area of a rectangle, one side of which is the E.M.F. and the other the projection of the current upon the E.M.F. line. Or, we may project the E.M.F. on

Fig. 33.

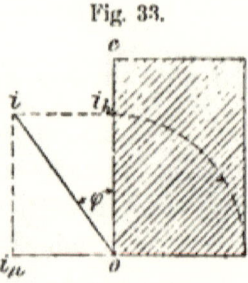

the current line, and thus form the rectangle. In measuring the surface it is of course necessary to bear in mind the scale to which E.M.F. and current have been plotted. If, for instance, the scale is 1 m.m. per volt and 1 m.m. per ampère, then the number of square m.m. contained in the rectangle is the number of watts. If, however, we plot the current to a scale of 1 m.m. per ampère, and the E.M.F. to a scale of 1 m.m. per 100 volts, then each square m.m. of surface represents 100 watts. In Fig. 33, Oi is the effective current and Oe the effective E.M.F. which leads by the angle ϕ. The shaded rectangle gives the effective power which is equivalent with that represented by a current i_h flowing under a potential difference e, but having no lag. We may thus assume the real current i replaced by two currents i_h and i_μ, between which there is a phase difference of 90°. These can be regarded as the components of the real current, the component i_h conveying all the power, and being therefore called the watt-component of the current, whilst i_μ conveys no power, and is called the wattless component, or briefly the idle current.

Combination of currents and pressures.—We have in the above an adaptation of the parallelogram of forces to the combination of currents of different phase; and this leads

Fig. 34.

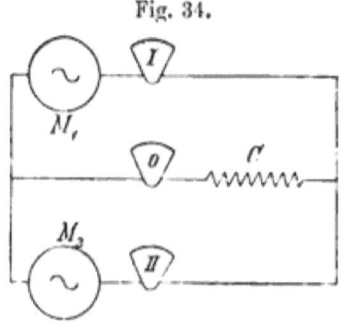

to the question whether such combinations are generally admissible, provided the currents are sine functions of the time and have the same frequency. Let us assume that the currents are produced by the alternators M_1 and M_2 (Fig. 34), which are mechanically geared so as to insure equality of phase and a constant lag. The machine currents are measured on the ampèremeters I and II, whilst the resulting current is measured on the ampèremeter O. The problem is to predetermine the current passing through the conductor C, if the machine currents and their phase difference are known. Let, in the clock diagram (Fig. 35), I' and I'' represent the maximum of the

machine currents as regards strength and relative phase.
At the moment to which the diagram refers, the alternator
M_1 gives the current Oi' and the alternator M_2 the current
Oi'', so that the total current passing through C is Oi' +
Oi''. If we draw the parallelogram $OI' II''$ we see at a
glance that the vertical distance between the points I and
I' is equal to the height of point I'' over the horizontal.
In other words the distance Oi is equal to the sum of Oi''
and Oi'; so that Oi represents correctly the strength of
the resultant current at the moment for which the clock
diagram has been drawn. The length Oi is nothing else

Fig. 35.

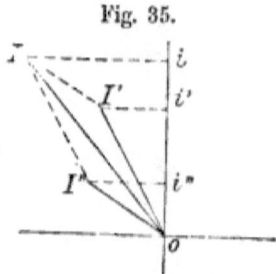

than the projection of OI upon the vertical, and since the
above reasoning holds good for any position of the vectors,
it is clear that the projection of OI will at all times give
the instantaneous value of the resultant current. We may
therefore imagine that the conductor C is traversed by a
single current the maximum value of which is graphically
represented by the resultant of the two current vectors I'
and I'', and the phase of which lies between the phases
of these two currents. If we further imagine all lengths
in the diagram reduced in the ratio $\sqrt{2} : 1$ there will be
no change in the angles, nor in the ratios of vectors, but
the resultant vector will then represent not the maximum,
but the effective value of the resultant current. It will be

clear that the above method of combining currents can be
applied to more than two currents. We first find the
resultant of two currents, then combine this resultant
with the third current, and obtain a new resultant, and so
on. It is not necessary in this operation to draw out the
various parallelograms of currents; all we need do is to
add the currents graphically after the manner of the
polygon of forces. The last line which closes the polygon

Fig. 36.

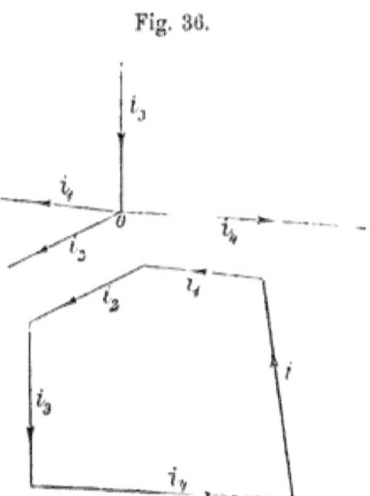

represents in magnitude and phase the resultant current.
Let, in Fig. 36, the lines i_1 to i_4 represent four currents
as regards phasal position, direction, and magnitude, then
by drawing the polygon of 5 sides, of which 4 correspond
to the current vectors, we obtain in the fifth side the
vector of the resultant current as regards phase, direction,
and magnitude.

Electromotive forces of the same frequency, but differing
in phase and magnitude, may be combined in the same
manner.

Let, in Fig. 37, M_1 and M_2 represent two alternators of
the same type, and mechanically coupled together so as to
insure equal frequency. Let three voltmeters O, I, and II,
be connected as shown, then I will indicate the terminal
pressure of M_1, II that of M_2, and O will show the resultant
pressure. The latter is not necessarily equal to the alge-
braic sum of the two other readings, but will as a rule be
smaller. It depends on the magnitude of the component
pressures and their phasal difference. After what has
already been said with regard to the combination of currents,
we need not explain the combination of electromotive

Fig. 37.   Fig. 38.

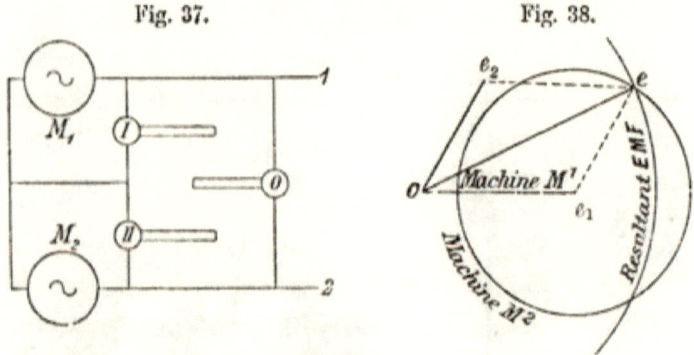

forces at length. If the three voltmeter readings are
available we can use them to determine the difference of
phase in the E.M. Forces of the two machines, as will
easily be understood from Fig. 38. Let Or_1 be the
voltage indicated on I, and draw round r_1 as centre a
circle with radius equal to the voltage indicated by II.
The resultant pressure must be a line joining O with
some as yet unknown point on the circle. To find this
point we need only describe a circle round O as centre with
a radius equal to the reading on the voltmeter O. The two
circles have two points of intersection, either of which may

be the end of the vector of resultant E.M.F. If the phase of M_1 is in advance over that of M_2 then the vector of M_2 will lie behind and therefore above that of M_1, the rotation of vectors being clock-wise. The resultant E.M.F. will then have the position shown in the diagram, and the angle of lag is $e \ Oe_2$.

Determination of the no load current.—In the beginning of this chapter we have considered the general principles for the determination of the power of an alternating current. The practical methods for measuring power will be given later on, as also the extension of these methods for the measurement of currents of non-sinusoidal form of wave. The investigation as far as it has been carried up to the present is, however, sufficient to enable us to determine the currents which the primary of a transformer takes when the secondary is open, and we now proceed to study this subject, which is of great practical importance. The importance lies in this, that transformers used for lighting are generally only loaded during very few hours daily, and if each were to take a large no load current, the ampère output of the central station might remain considerable even during those hours when very little or no light is required. To make this matter clear, let us assume a central station for the supply of 100,000 50-watt lamps installed. We shall then require a number of transformers with a total output of 5000 Kwt., since the case may arise that one or other of the users lights all the lamps installed in his house at the same time. The total output of the alternators, however, need not be as large as 5000 Kwt. Experience has shown that of all the lamps connected to a central station only some 30 to 70 per cent. are ever lit simultaneously, the exact percentage depending on the character of the locality served from the station. Assuming 60 per cent. as an example, we would have to supply alternators for a total supply of 3000 Kwt. to the lamps. The day supply is of course very small, and may be taken as from 3 to 4 per cent. Assuming $3\frac{1}{2}$ per cent. as an average, we should have to provide a little over 100 Kwt. during the day-time. To this output must be added the loss in hysteresis, for which we may take 2 per cent. of the total transformer output, or say another 100 Kwt., so that an alternator of 200 Kwt. output ought to suffice. In reality, however, a larger alternator will be required, because more current than corresponds to the power is taken by the transformers, the discrepancy being the greater the larger the no load current of each transformer. With a no load current of 10 per cent., the current output of the central station, apart from the true power output of 200 Kwt., would amount to an apparent power of 500 Kwt., and with a no load current of 5 per cent. the current output would correspond to an apparent power of 250 Kwt. It is thus clear that in order to reduce as far as possible the amount of machinery which must be kept going during the day-time, the transformers must be so constructed as to require a minimum of no load current.

The no load current is required for magnetizing the iron to that value of B which corresponds to the primary E.M.F., and for supplying the power lost through hysteresis. The magnetizing current can be calculated according to the well-known laws of electro-magnets, when the dimensions of the iron part, the quality of the iron, and the number of turns of primary wire are known.

Let l in Fig. 39 represent the mean length of the path of the magnetic flux as determined from the drawing, and μ the permeability of the particular quality of iron used at the induction B, then the magnetizing force of the current I_μ passing through n turns is $4 \pi n I_\mu : l$, where

Fig. 39.

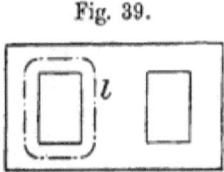

I_μ is taken in absolute measure. If I_μ be taken in ampères the magnetizing force becomes $0\cdot4\,\pi\,n\,I_\mu : l$ and the induction produced is—

$$B = \mu \frac{0\cdot4\,\pi\,n\,I_\mu}{l}$$

I_μ being the maximum value of the current wave. If i_μ represents the effective value, we have $I_\mu = \sqrt{2}\,i_\mu$ and—

$$B = \mu \frac{1\cdot78\,n\,i_\mu}{l}$$

$$i_\mu = \frac{B\,l}{1\cdot78\,\mu\,n}$$

The change in the magnetic flux produces in the coil an E.M.F., the phase of which is $90°$ behind that of the current i_μ, as is easily seen from the following. If the current has attained its maximum, B is also a maximum, and the E.M.F. is therefore zero. If the current and therefore the flux pass through zero, the E.M.F. is a maximum. To the maximum of current corresponds the E.M.F. zero and *vice versâ ;* a relation which can only exist if the angle ϕ in Fig. 31 is $90°$. Then $e\,i \cos\phi = 0 ;$ there is no power given out by the current. We thus find that the magnetizing current i_μ carries no power; its vector in the clock diagram (Fig. 33) stands at right

angles to the vector of the E.M.F. as shown in the
diagram.

Power is, however, required to cover the hysteresis
loss; and the component of the no load current which
carries this power must coincide in phase with the E.M.F.
In Fig. 33 this component is i_h. The total no load
current is therefore the resultant of two currents, namely—

The wattless magnetizing current i_μ.

The watt current required to cover the hysteresis loss i_h.

Since these components are at right angles to each other,
we can calculate the resultant no load current $i\,o$ from
the equation—

$$i_o = \sqrt{i_\mu{}^2 + i_h{}^2}$$

Up to the present we have assumed that the magnetizing
force is solely required to drive the flux through iron; in
other words, that the path of lines is either not interrupted by joints, or that if joints there be, they have no
magnetic resistance. This condition is fulfilled with over-
lapping joints, such as are shown in Figs. 18—22; it is,
however, not fulfilled in but-joints, as shown in Figs. 16
and 25. If for the convenient fitting together but-joints
are used, it is obvious that at every joint a certain
magnetic resistance is introduced, since, for reasons already
stated in Chapter III., no perfect junction face to face can
be obtained nor permitted, even if it were mechanically
possible. We must therefore take account of the fact
that the faces are separated by a certain distance, and
that the lines of force have to pass through a non-magnetic
medium at every joint. Let δ be the combined thickness
of these non-magnetic layers (that is $\dfrac{\delta}{2}$ or $\dfrac{\delta}{4}$ the thickness of each layer with two and four joints respectively),

and B represent the induction at the joint, then, since $\mu = 1$, the magnetizing force required for overcoming the resistance of the joints is—

$$0\cdot4 \, \pi \, n \, I = \delta \, B$$

and the effective ampère-turns required are—

$$i \, n = \frac{B \, \delta}{1\cdot78}$$

The current i must be added to the magnetizing current previously determined, and we thus obtain the magnetizing current for a transformer with but-joints of total thickness δ by the equation—

$$i_\mu = \frac{B}{1\cdot78 \, n} \left(\frac{l}{\mu} + \delta \right)$$

l and δ being given in centimeter. If there are overlapping joints or no joints at all $\delta = o$, and we obtain the expression previously given.

If for the moment we confine ourselves to transformers having no but-joints, we are able to utilize the formulæ here given for the determination of the permeability by experimenting with finished transformers. For this purpose the transformer is worked with open secondary, whilst the primary current and power supplied are measured. It is convenient to use for this experiment the low-pressure winding as the primary. The E.M.F. and frequency are also observed, and from these data can be calculated the induction B and the component i_h required to cover the loss of power through hysteresis. If P_v is the loss and c the supply pressure, we have—

$$i_h = P_v : c$$

With good, and even with middling good, transformers with closed magnetic circuit the no load current is so small that loss of power by ohmic resistance may be neglected;

hence P_v is simply the power measured on open circuit. Let i_o be the no load current observed, then the magnetizing current can be calculated from the formula—

$$i_\mu = \sqrt{i_o^2 - i_h^2}$$

and the permeability from the formula—

$$\mu = \frac{B\,l}{1,78\,n\,i_\mu}$$

If the experiment be repeated for different values of primary E.M.F., we are thus able to find a series of corresponding values for B and μ, and the results obtained may then be used in designing new transformers.

The following table gives corresponding values of B and μ which were found experimentally in the manner above explained. The iron contained in the transformers under test was of different quality, but lay, as regards hysteresis, between the limits indicated by the two curves in Fig. 9. For the rest the experiments showed that the permeability cannot be taken as a measure for the value of the iron as regards low hysteresis loss, and that the difference in permeability for different brands of transformer-plates are generally small. The values here given are averages.

$B =$	2000	3000	4000	5000	6000	7000
$M =$	1300	1720	2070	2330	2570	2780

The dotted curve in Fig. 9 represents these values graphically.

Influence of but-joints.—It remains yet to investigate what effect but-joints produce as regards the no load current. On that component of it which is required for covering the hysteresis and eddy current losses, they have, of course, no influence, since these losses do not depend on the magnetic resistance. The wattless component of the no load current, being directly proportional to the magnetic resistance, is, however, strongly influenced by the presence of but-joints. To show this we may apply the formula—

$$i_\mu = \frac{B}{1,78\,n}\left(\frac{l}{\mu} + \delta\right)$$

to a practical example. In core transformers of the types Figs. 15 or 16 we should have four but-joints. The gap in each joint can, even with the most careful workmanship and the use of specially tough material as an insertion, not be made smaller than 0·5 millimeter. We thus obtain $\delta = 0·2$ centimeter. The permeability is from Fig. 9 of the order 2000. The mean length of magnetic path varies of course with the size of the transformer. With small transformers of 1 to 10 Kwt. it may be taken as between 70 and 160 centimeter; with large transformers of 100 Kwt. or thereabouts, l would be of the order 300 centimeter. Assuming as a fair average for l 100 c.m. in small and 250 c.m. in large transformers, then the fraction $l : \mu$ will lie between the limits 0·05 and 0·15. The term in brackets in the equation for i_μ will therefore be increased from 0·05 to 0·25 in small, and from 0·15 to 0·35 in large transformers. This is an increase of 400 and 133 per cent. respectively.

If in a small transformer without but-joints the magnetizing current is 4 per cent. and the hysteresis current 3 per cent. of the full load current, the no load current will be—

$$i_0 = \sqrt{3^2 + 4^2} = 5 \text{ per cent.}$$

Now let us build the same transformer with but-joints. This will not alter the hysteresis current, which remains

3 per cent. as before, but the magnetizing current will be increased to 20 per cent., giving a no load current of—

$$i_o = \sqrt{3^2 + 20^2} = 20{\cdot}2 \text{ per cent.}$$

Let in a large transformer without but-joints

$i_h = 1{\cdot}5\,\%$ and $i_\mu = 2\,\%$, then $i_o = \sqrt{1{\cdot}5^2 + 2^2} = 2{\cdot}5\,\%$. The same transformer built with but-joints would have $i_\mu = 4{\cdot}6\,\%$ and $i_o = \sqrt{1{\cdot}5^2 + 4{\cdot}6^2} = 4{\cdot}85\,\%$.

By using but-joints we have thus in the small transformer quadrupled, and in the large transformer nearly doubled, the no load current.

Our examples referred to core transformers. With shell transformers the use of but-joints would be still more objectionable, since the length of the magnetic path is only about one-third as compared with core transformers, and the influence of an increase in magnetic resistance correspondingly greater. For this reason shell transformers are never built with but-joints, but always with over-lapping joints according to one or other of the methods described in Chapter III. For large core transformers but-joints may be permitted, partly because their influence is less felt than in small transformers, but chiefly because large transformers are nearly always working with some load, and then the influence of a larger or smaller magnetizing current is hardly felt. Small transformers, especially those for private house lighting from a central station, should never be made with but-joints, as they have long periods of no load during which they burden the station with a large current output out of all proportion with the useful power output.

CHAPTER V

Construction of a transformer.—In order to demonstrate
the practical application of the rules and formulæ given
in the previous chapters, we now proceed to work out
a design for a particular transformer, with a transforming ratio of 2000 : 100. For this purpose we select a
core transformer of the type Fig. 15, and assume the
thickness of core to be 125 m.m. The winding space
thus becomes by our formulæ $a = 160$ m.m ; $b = 450$ m.m.
Let the frequency be $\sim = 50$. To start with, let us
assume $B = 5000$, with the reservation, that should it be
found advantageous we will alter the induction. The
core and yoke-plates are held together by insulated bolts,
as already explained, and the coils are wound on separate
cylinders of paper, or preferably micanite, so that the
windings may be independently prepared and slipped on.
It is immaterial which winding is inside and which outside ; we shall place the low pressure or secondary winding next to the core, and the high-pressure winding
outside.

To save copper, we chamfer the edges of the core by
say 20 m.m., as shown in Fig. 40. This is done by

stepping the width of the outer plates. After all the plates are assembled the core is bound up by strong tape to keep the plates from bulging. The thickness of this tape serving is about 2 m.m. In order to enable the low-pressure coil to be easily slipped on, we require a clearance of 2 m.m. over the corners, so that the inner diameter of the supporting cylinder becomes 160 m.m. Its thickness need not exceed 5 m.m., bringing the inner diameter of the coil itself to 170 m.m. The depth of winding and the

Fig. 40.

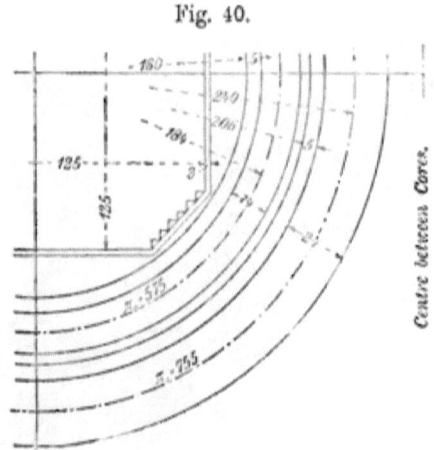

mean perimeter can at present only be fixed approximately from the following consideration. The dimensions of the iron part being given, we find the distance of centres of cores to be $125 + 160 = 285$ m.m. The external diameter of the primary winding can therefore not exceed 285 m. m., and must in reality be somewhat smaller, since a certain clearance is necessary not only to avoid touching of the two coils, but also to make up for possible irregularities in the manufacture, and leave a margin in case the same size of transformer should have at any future time to

be made for a higher primary pressure, when finer wire requiring more space would have to be used. A certain
clearance is also necessary to allow for the circulation of
air or oil, whereby the coils are kept from over-heating.
If we fix this clearance at 20 m.m., we obtain for the
external diameter of each primary coil 265 m.m., so that
between the inner surface of the secondary and the outer
surface of the primary winding, there remains a distance
of $\frac{1}{2}$ (265 – 170) = 47·5 m.m. This distance is made up of
a, the depth of winding in the secondary coil; b, the
clearance between the latter and the inner side of the
primary cylinder; c, the thickness of the latter; and d, the
depth of the primary winding.

For clearance we may reckon 4 m.m., and for the thickness of primary cylinder 5 m.m., leaving 47·5 – 9 = 38·5
m.m., to be apportioned between the two windings. It is
obvious that the depth of primary winding will be more
than half this amount, because not only is more space
required for the insulation of the thinner wire, but the
perimeter being larger the gauge of the wire must be
correspondingly increased in order to avoid getting too
much resistance in the primary. As a first attempt in
apportioning the space we may assume that 60 per cent.
of it will be filled by the primary, and 40 per cent. by the
secondary winding.

Best distribution of copper.—The correct distribution of
the available space between the two windings is in so far
important as the total loss by ohmic resistance becomes a
minimum if it is equally divided between the two windings.
This may be shown as follows. Let the total winding
space available for both coils be b (in our case $b = 38·5$
m.m.), and the depth of the secondary winding a, then $b - a$
is the depth of the primary winding. Let D be the inner

diameter of the secondary coil. Its resistance is proportional to the number of turns n_2 and to the mean perimeter of one turn $\pi(D + a)$; it is inversely proportional to the cross section of wire; or to put it in another way, the resistance is with a given length of wire inversely proportional to the depth of winding, a. By comprising all constants into one co-efficient K, we have for the loss in the secondary coil—

$$P_{v2} = K n_2 \frac{D + a}{a} i_2{}^2$$

For the primary coil we obtain a similar expression, except that the co-efficient K must be multiplied by the ratio of windings $n_1 n_2$, and we thus obtain—

$$P_{v1} = K n_1 \frac{D + a + b}{b - a} \frac{n_1}{n_2} i_1{}^2$$

Since $n_1{}^2 i_1{}^2 = n_2{}^2 i_2{}^2$, we can also write—

$$P_{v1} = K n_2 \frac{D + a + b}{b - a} i_2{}^2$$

The total loss by ohmic resistance is therefore—

$$P_v = K n_2 i_2{}^2 \left(\frac{D + a}{a} + \frac{D + a + b}{b - a} \right)$$

In order to make this loss a minimum, a must be so chosen that the term in brackets becomes a minimum. By differentiating and equating to zero, we obtain after some reduction—

$$a^2 + a D - \frac{bD}{2} = 0$$

$$a = -\frac{D}{2} + \sqrt{\frac{D^2}{4} + \frac{bD}{2}}$$

Since a negative depth of winding is physically impossible, we take the upper sign—

$$a = -\frac{D}{2} + \sqrt{\frac{D^2}{4} + \frac{bD}{2}}$$

By equalling the two terms in brackets we also get—

$$a^2 + a\,D - \frac{bD}{2} = 0$$

and we see from this that there will be minimum loss by ohmic resistance if the loss is equally divided between the two circuits. If therefore our first assumption regarding the depth of winding does not produce equality of losses, we must redistribute the available winding space accordingly so as to satisfy the above equation for a. We design for $B = 5000$ and $\sim\, = 50$ the windings, retaining for the present the ratio of 40 to 60 per cent. for secondary and primary winding respectively. On calculating the ohmic losses, we find that the primary loses too much and the secondary too little power. We therefore reduce the depth of winding in the secondary, and increase it in the primary. It is superfluous to give the calculation in detail; the result is—

Best depth of winding for the secondary   14 m.m.

    ”   ”   ”   ”   ”   primary   24 m.m.

We may now draw the coils, and determine from the drawing the exact mean perimeter of each. This gives—

$$\pi_2 = 0{,}575 \text{ m.} \qquad \pi_1 = 0{,}755 \text{ m.}$$

The next step in the design is the determination of the hysteresis loss from the weight of iron, the induction, and the frequency. Assuming that we use best transformer plates, the lower of the two curves in Fig. 9 is to be taken. According to this curve the loss of power with $B = 5000$, and $\sim\, = 100$ is 1·8 watt per kilogram, and since our frequency is $\sim\, = 50$, we have a loss of 0·9 watt per kilogram. The net section of iron in the cores is 130 square c.m., and in the yokes which are not chamfered 136 square c.m. The weights can be calculated from the dimensions, and we thus obtain for the total hysteresis loss—

2 Cores 116·8 kg. B = 5000 at 0·9 watt ...	105 watt
2 Yokes 61·2 „ B = 4770 at 0·85 „ ...	52 „
Total weight 179 kg.	Total hysteresis loss 157 watt

We next determine from the drawing the total cooling surface with 12,000 square centimeters. Assuming for the present that the transformer is to be worked in a case without oil, and that its temperature rise is to be 54° C., we find from the curve for air Fig. 29 σ = 37, and can now determine the load by fixing the total losses at 12,000 : 37 = 324 watt. The copper heat may therefore amount to 324 − 157 = 167 watt, or with the best distribution of copper to 83·5 watt for each winding.

The next step is to more accurately determine the windings. In doing so it is important to observe that besides the loss of pressure due to ohmic resistance, there is also loss of pressure due to magnetic leakage. The latter cannot be calculated, but may be closely estimated from experiments with other transformers of the same type. In our design, where the coils are not placed upon, but within, each other, the loss of pressure due to leakage and with a noninductive load (such as glow lamps) is very small, and may be roughly taken at 1 per cent. If we allow an ohmic loss of pressure of $1\frac{1}{2}$ per cent., we have a total loss of $2\frac{1}{2}$ per cent., that is to say, the terminal pressure on the secondary will be $2\frac{1}{2}$ per cent. higher at no load than at full load.

Since N = 130 × 5000 = 650,000, and $\sim$ = 50, we obtain from the formula—

$$E_2 = 4\cdot44 \sim n_2 N\ 10^{-\wedge}$$

the number of turns n_2 in the secondary winding. This
must obviously be a whole number, and if we wish to get
the winding symmetrical on both limbs (so as to fully
utilize the available space) it must also be an even number.
The nearest number which satisfies these conditions is

$$n_2 = 70,$$

whereby for $B = 5000$ $E_2 = 101\cdot23$.

If there were no appreciable voltage drop either through
ohmic resistance or magnetic leakage, a condition fulfilled
when the transformer is working on open circuit, then the
number of primary turns would be $2000 : 100 = 20$ times
as great as the number of secondary turns. This would
give $n_1 = 1400$. With 2000 volts on the primary terminals, the pressure on the secondary terminals would at no
load be exactly 100 volts, but at full load $2\frac{1}{2}$ per cent. loss,
or only $97\cdot5$ volts. If then we wish to have 100 volts
pressure at full load, we must alter the transforming ratio
at no load by $2\frac{1}{2}$ per cent., that is to say, we must reduce

the primary winding by $\dfrac{2\cdot5 \times 1400}{100} = 35$ turns. We have
then

$$n_1 = 1365$$

On open circuit E_2 will then not be $101\cdot23$ volts, but
$102\cdot5$ volts, and the induction will increase in the same
ratio, namely, by $102\cdot5 - 101\cdot23 = 1\cdot27$ per cent. It is now
not 5000 but 5063, and the hysteresis loss will be increased
from 157 to 161 watt.

The cross section of wire may now be determined. For
fixing the length of the coils we have to consider the height
of the window in the iron frame (in our case 45 c.m.), and
leave sufficient space for clearance and the end flanges of

the cylinders. The total space required for these purposes is about 3½ c.m., leaving 41·5 c.m. net length of coil. Each secondary coil must contain 35 turns of wire. If these were arranged in a single layer, the wire would have to be wound on edge. Although this presents no difficulty with naked wire which is afterwards insulated by fibre insertion, it is not so easy with cotton-covered wire, and in this case it would be better to wind the wire on the flat and make two layers, one with 18 and the other with 17 turns. Since the space of one turn is lost in crossing over from the lower to the upper layer, we must arrange the width of the wire to be not $\frac{1}{18}$th, but $\frac{1}{19}$th of the net winding space. This gives 415/19 = 21·8 m.m. The thickness of the wire is already determined by the depth of winding, which we found must be 14 m.m. Allowing 0·5 m.m. for the thickness of covering (or 1 m.m. in all) we find that the section of the wire will be 6 × 20·8 m.m. Since it is, however, scarcely possible to lay on succeeding turns with mathematical accuracy, it will be advisable to take the width a little less, say 20 m.m., so that the actual cross-section of the wire becomes 6 × 20 = 120 square m.m. The length of winding is 70 × 0·575 = 40·5 m., and if we allow 0·5 m. for connections, we can take 41 m. as the basis on which to calculate the resistance of the secondary winding. The formula for the resistance, taking an average rise of temperature into account, is—

$$W_2 = \frac{0\cdot 02 \; l_2}{q}$$

l_2 being the length in meter and q the cross-section in square millimeter. We thus obtain—

$$W_2 = 0\cdot 00682$$

A similar calculation made for the primary winding shows

that we have to use round wire of 3·1 m.m. diameter (covered to 3·67 m.m.) in six layers of 122 turns, and one layer of ten turns on one and eleven turns on the other limb. The length of wire is—

$$l_1 = 1030 \text{ m.}$$

and its resistance warm is

$$W_1 = 2·8$$

We have now all the data required for calculating the losses at different loads. For convenience they are given in the following table—

Load in kilowatts	8	9	10	11	12	13	14	15
Secondary current ampère	80	90	100	110	120	130	140	150
Primary current ,,	4·125	4·631	5·150	5·661	6·180	6·70	7·22	7·44
Copper heat watt ...	91	115	143	172	205	241	278	321
Hysteresis ,, ...	161	161	161	161	161	161	161	161
Total losses ,, ...	252	276	304	333	366	402	439	482
Percentage loss ...	3·05	3·06	3·04	3·03	3·06	3·09	3·15	3·22
Cooling surface σ ...	47·6	43·5	39·5	36·0	32·8	30·0	27·4	25·0
Temperature rise in air ...	45·5	48·5	51·3	54·0	57·2	60·6	64	67
Temperature rise in oil ...	34	36·0	38·5	40·6	43	45	47	49·3

If the transformer is not to be cooled by oil, we see from this table that the maximum load it can carry with the heating limit assumed is 11 Kwt. The total loss is then 333 watt, so that 11,333 watt must be supplied to the primary, making the efficiency—

$$\eta = \frac{11,000}{11,333} = 97 \text{ per cent.}$$

If we use oil we may load the same transformer to 15 Kwt., and obtain approximately the same efficiency.

Cost of active material.—The efficiency is however not the only guide in judging the merit of a design. We must also take into consideration at what cost in material the result is obtained. Assuming the finished or cut plates to cost 9*d.* per kg., and the insulated copper wire 1*s.* 9*d.* per kg. (4*d.* and 9½*d.* per lb. respectively), then we have—

Iron	179	kg. at	9*d.*	. . . . . .	134*s.*
Copper	111·5	kg. at	1*s.* 9*d.*	. . . . . .	195*s.*
Total weight	290·5		Cost of active material 329*s.*		

The merit of the transformer worked in air and oil is then shown by the following figures—

	IN AIR.	IN OIL.
Output in Kwt.	11	15
Weight of active material per Kwt.	26·4 kg.	19·4 kg.
Cost of active material per Kwt.	29·9*s.*	21·9*s.*

Best distribution of losses.—If we use oil filling for the transformer case we obtain a lighter and cheaper apparatus per Kwt. output than if we leave the case empty. On the other hand, the percentage loss is a trifle larger in the transformer working in oil, namely, 3·22 per cent. against 3·03 per cent. for the transformer working in air. The efficiency is a maximum with a load of 11 Kwt., when the hysteresis loss is about the same as the ohmic loss. If the output is less than 11 Kwt. we have a lower efficiency, whilst the hysteresis loss is larger than the ohmic loss; and if we increase the output beyond 11 Kwt. we also lower the efficiency, but now the ohmic loss is larger than the hysteresis loss. These facts let it appear probable that we obtain not only in this particular transformer, but generally in all transformers, a maximum of efficiency at that load at which the hysteresis loss and the ohmic loss are equal. If this assumption is correct, then the design of our transformer when used with oil for 15 Kwt. output, must be capable of improvement by increasing the hysteresis loss and decreasing the ohmic loss. The total loss in the present design is 482 watt. We shall now alter the windings so as to increase the induction by such an amount as will bring the hysteresis loss up to about half this figure, or 241 watt. Retaining the same core, we may now allow a loss of 1·34 watt. per kg., or 2·68 watt. at $\sim = 100$, to which corresponds by the lower curve in Fig. 9, $B = 6300$.

To this induction corresponds $n_2 = 56·5$; since it is however impossible to arrange for a fraction of a turn, and since for the sake of symmetry we must have an even number of turns, we make $n_2 = 56$. This gives $N = 824,000$ and $B = 6350$.

The hysteresis loss can now be calculated more exactly as follows—

2 Cores 116·8 kg. $B = 6350$ at 1·35 watt. ... 158 watt.
2 Yokes 61·2 „ $B = 6080$ „ 1·25 „ ... 77 „
Total 235 „

Owing to the reduction in the number of turns we may now make the secondary wire 26 × 6 m.m. The resistance warm will be 0·00415 ohm. The primary will have 20 × 56 less 2½ per cent. = 1092 turns. There is room for round wire of 3·5 m.m. diameter (covered to 4·2 m.m.). Of this wire 98 turns go to one layer, so that we shall require five layers and 54 turns on each limb. The resistance warm is 1·72 ohm. The weight of copper in the secondary is 44·5 kg. and in the primary 70·5 kg.; total 115 kg.

The following table gives the condition of working at various loads—

Load in Kwt.	...	8	12	15	16	17
Secondary current	...	80	120	150	160	170
Primary current	...	4·15	6·18	7·72	8·23	8·75
Ohmic loss	...	56	115	195	222	252
Hysteresis loss	...	235	235	235	235	235
Total loss	...	291	350	430	457	487
Percentage loss	...	3·64	2·92	2·87	2·86	2·86

We see from this table that the least percentage loss, and therefore the highest efficiency, is obtained at that load at which the ohmic loss is approximately equal to the hysteresis loss. The weight of active material is in this transformer 294 kg., and its cost determined on the same basis as before is 335s. Since the load may be 16 Kwt. we have 18·4 kg. and 20·9s. per Kwt. output. This is an improvement on the previous design brought about by the heavier magnetic stress on the iron which was rendered possible through cooling with oil.

Let us now reduce the linear dimensions so far that with $B = 6350$ the output is reduced from 16 to 15 Kwt.

The ratio of reduction is $\sqrt[3]{\dfrac{15}{16}}$ for the linear dimensions, and $15 : 16$ for the losses. The object of this reduction is to bring both transformers (namely, that in which $B = 5000$, and that in which $B = 6350$) to the same output of 15 Kwt., in order to have the same basis of comparison. Both transformers are cooled by oil.

DESIGN.					I.	II.
Induction	...	...	...	...	5000	6350
Output Kwt.	...	...	...	...	15	15
Temperature rise degrees centigrade		...		...	49·3	49
Hysteresis loss	...	...	...	...	161	220
Ohmic loss	...	...	...	...	321	208

DESIGN.					I.	II.
Total loss	...	...	...	...	482	428
Percentage loss	...	...	...	...	3 22	2·85
Weight per Kwt. output kg.		...	...		19·4	18·4
Cost per Kwt. output 6s.			...	...	21·9	20·9

Economy in working.—These figures show clearly that design II. is preferable to design I. both as regards first cost and economy in working, provided the transformer is required to work permanently under full load. This is generally the case in power transmission, for which purposes transformers should therefore be so designed that the hysteresis loss is approximately equal to the ohmic loss. The case is however different with transformers required chiefly for lighting. The average burning time of all lamps installed is of course much shorter than the time during which the transformer must be kept at work. The transformer must be large enough to feed all the lamps installed, as it may occasionally happen that all are wanted at the same time, though as a rule the number of lamps burning will be smaller than the number of lamps installed. If a transformer for a single house is connected to a central station it is at work day and night, that is for 8760 hours during the year, whilst with an average burning time of 600 hours per lamp, the whole yearly output actually obtained is only $\frac{600}{8760}$ times the possible yearly output. The hysteresis loss goes on for 8760 hours; the ohmic loss only for the time that lamps are burning. The latter loss must be less than corresponds to full output during 600 hours, since this loss varies as the square of the current, and maximum current does not last the full 600 hours, but a much shorter time.

Assume a house with 100 lamps installed, and let the

ohmic loss in the transformer when all the lamps are in
use be 100 watt. If all the lamps were simultaneously
lighted and extinguished, then for a yearly burning time
of 600 hours the loss would be 60 Board of Trade units.
In reality the loss will be smaller, because a smaller
number of lamps will be used for a longer time than 600
hours, notwithstanding the fact that the total yearly lamp-
hours will be 60,000. If the time-table for the lamps
alight is as given below, the loss is reduced to 22·7 units.

LAMPS ALIGHT.	HOURS.	LAMP HOURS.	WATT.	LOSS IN UNITS.
100	50	5000	100	5
70	100	7000	49	4·9
40	500	20,000	16	8
20	1000	20,000	4	4
10	800	8000	1	0·8
	2450	60,000		22·7

The average burning time of all lamps is, as before
assumed, 600 hours; some of the lamps are however in use
for a much longer time, with the result that the ohmic
loss in the transformer is reduced by about 38 per cent. as
compared with the loss which would obtain if all lamps
were always used simultaneously. Applying the above
reasoning to the comparison of the two transformer designs
(I. and II.) when used for lighting, we come to a totally
different conclusion to that we reached when using the
transformers for power purposes.

Since 100 watt. maximum ohmic loss means a yearly
loss of 22·7 units, the ohmic loss of design I. will be
$22·7 \frac{321}{100} = 73$ units yearly; and that of design II. will be
$22·7 \frac{208}{100} = 47·3$ units. The hysteresis losses are for these
two designs, 1410 and 1930 units respectively.

		DESIGN.	
		I.	II.
B = ...		5000	6350
Annual ohmic loss in units	...	73	47·3
Annual hysteresis loss in units	...	1410	1930
Total annual loss in units	...	1483	1977
Annual output in units ...	...	9000	9000
Annual efficiency per cent.	...	86	82

As regards annual efficiency, that is the ratio of the units annually supplied to and obtained from the transformer, design I. is decidedly better than design II. On the other hand, the first cost of design I. is somewhat greater, namely 21·9s. against 20·9s. for active material alone. To this must be added the cost of other materials, the case, wages, shop and trade charges, and manufacturer's profit. As a fair average we may take it that the net selling price of the finished transformer amounts to 2½ or 3 times the cost of its active material. The net cost would therefore be for—

Design I.	...	...	...	...	£45.
Design II.	...	...	...	...	£43.

The difference in cost is only £2, and against this we have to set off a yearly waste of 494 units. Taking the engine-room charge for a unit at the central station at only 1d., the units wasted represent £2 1s. 1d.; or in other words, the more expensive design will pay for itself within one year. We may now formulate the results of our investigation as follows—

Transformers for power should be so designed that the stress in the iron is high and in the copper moderate; iron and copper losses should be approximately equal. Transformers for lighting should be so designed that the stress in the iron is small, and in the copper large. The

H

copper loss should be greater than the iron loss. It is obvious that this condition must not be pushed beyond the limits imposed by temperature rise and voltage drop.

Constructive details.—Figs. 41 to 44 give details of the design of the transformer here discussed. The output is 10 Kwt. with air cooling and 15 Kwt. with oil cooling. The

Fig. 41.

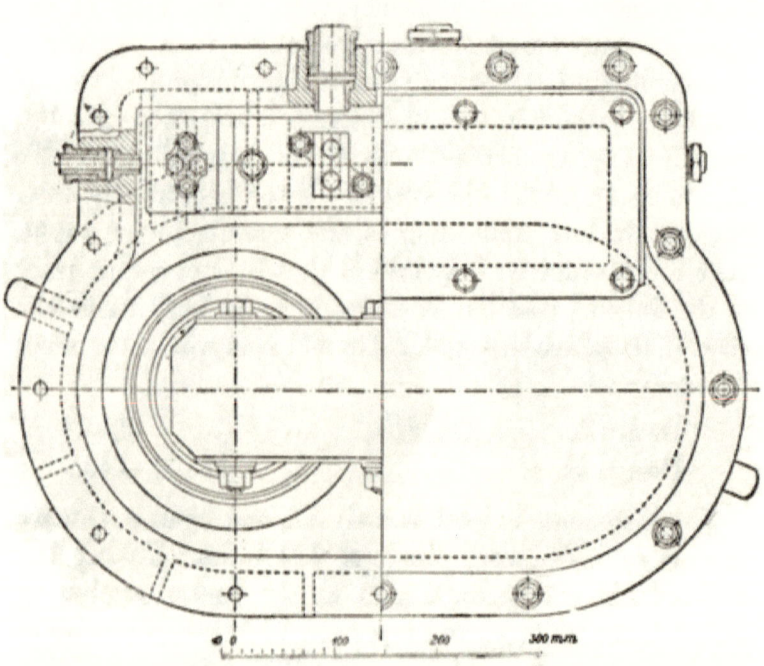

transformer is placed into a cast-iron case, and is therefore adapted for use in a cellar or other damp place, or in the open air in a damp climate. For outdoor use in a dry climate the case need not be completely closed, but need only be sufficient to protect the apparatus from rain, and for dry interior situations the case may be replaced by perforated covering (see Figs. 103, 110, 125). The advan-

tage of a perforated case is that the cooling effect of the
air is greater. It may be taken as represented by the
lower curve in Fig. 29.

The plates for the core and yoke are cut to size and
punched for the bolt-holes, then laid together with an
insertion of very thin paper. Some makers use varnish
instead of paper, but this is not so reliable an insulation.
In building up, the lower yoke and the two cores are first
made up, the coils are then inserted, and lastly the plates
of the top yoke are put in. The coils are wound on paper
cylinders, which at their lower ends are provided with
flanges to prevent the coils slipping. In winding the
coils it is advisable to wrap each layer with a sheet of
thin paraffined calico, which is doubled back at the ends
so as to give additional insulation between adjacent layers.

The thickness of the cotton covering on the wire depends
on its diameter (or equivalent diameter if rectangular wire
be used), the voltage, and the quality of the cotton and
number of coverings. There must at least be two coverings, though treble covering with very fine cotton is still
better. For very stout wires an additional braiding is advisable. For work up to 3000 volts the thickness of the
covering in millimetres should not be less than

$$\delta = 0\cdot13 + 0\cdot06 \, d,$$

when d is the diameter (or equivalent diameter) of the
naked wire in millimetres. The diameter of the covered
wire is then

$$d_1 = d + 2 \, \delta,$$
$$d_1 = 0\cdot26 + 0\cdot12 \, d.$$

Wire of large rectangular section may also be wound
naked, suitable strips of fibre or other insulating material
being wound in, or afterwards inserted.

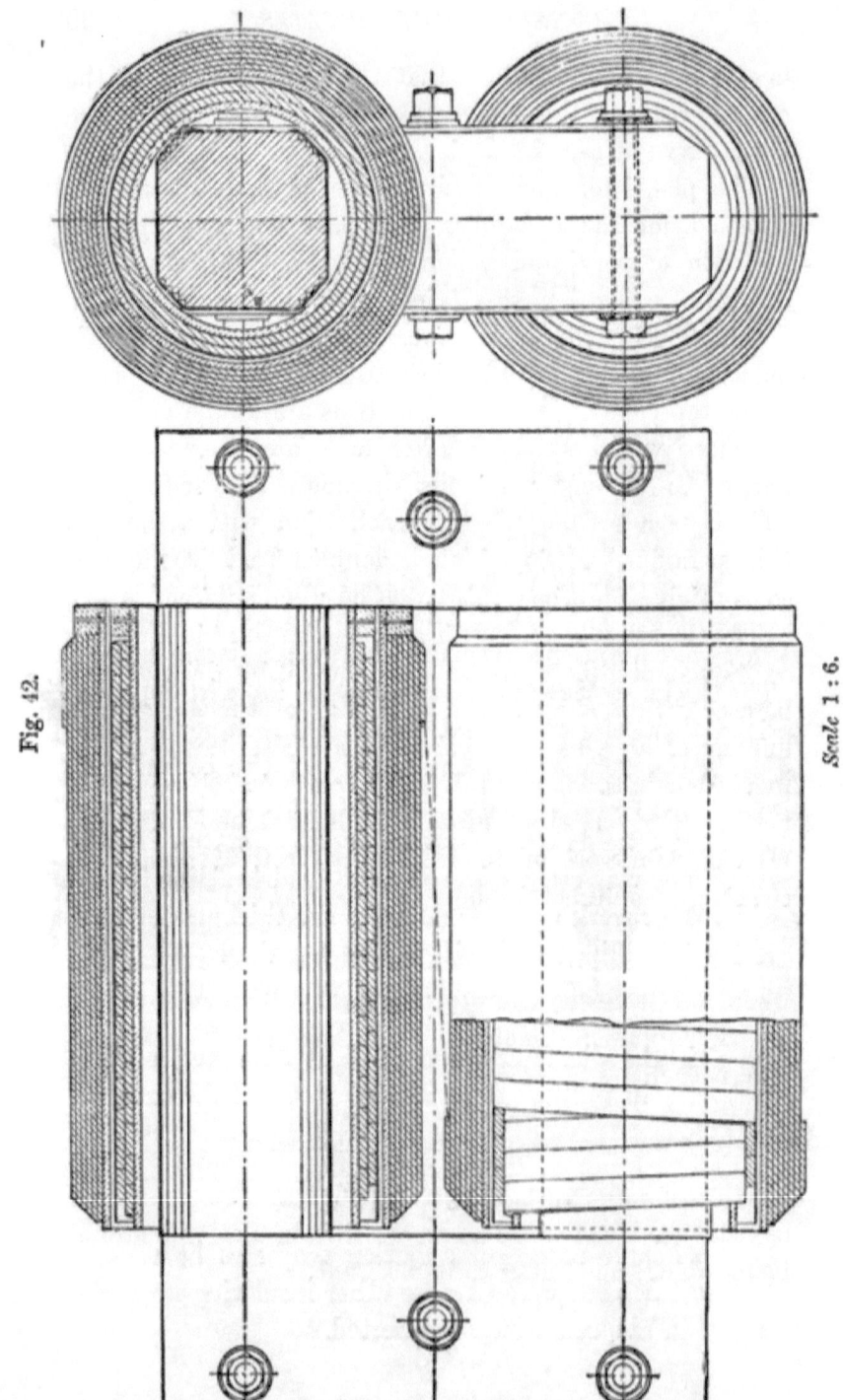

Fig. 12.

Scale 1 : 6.

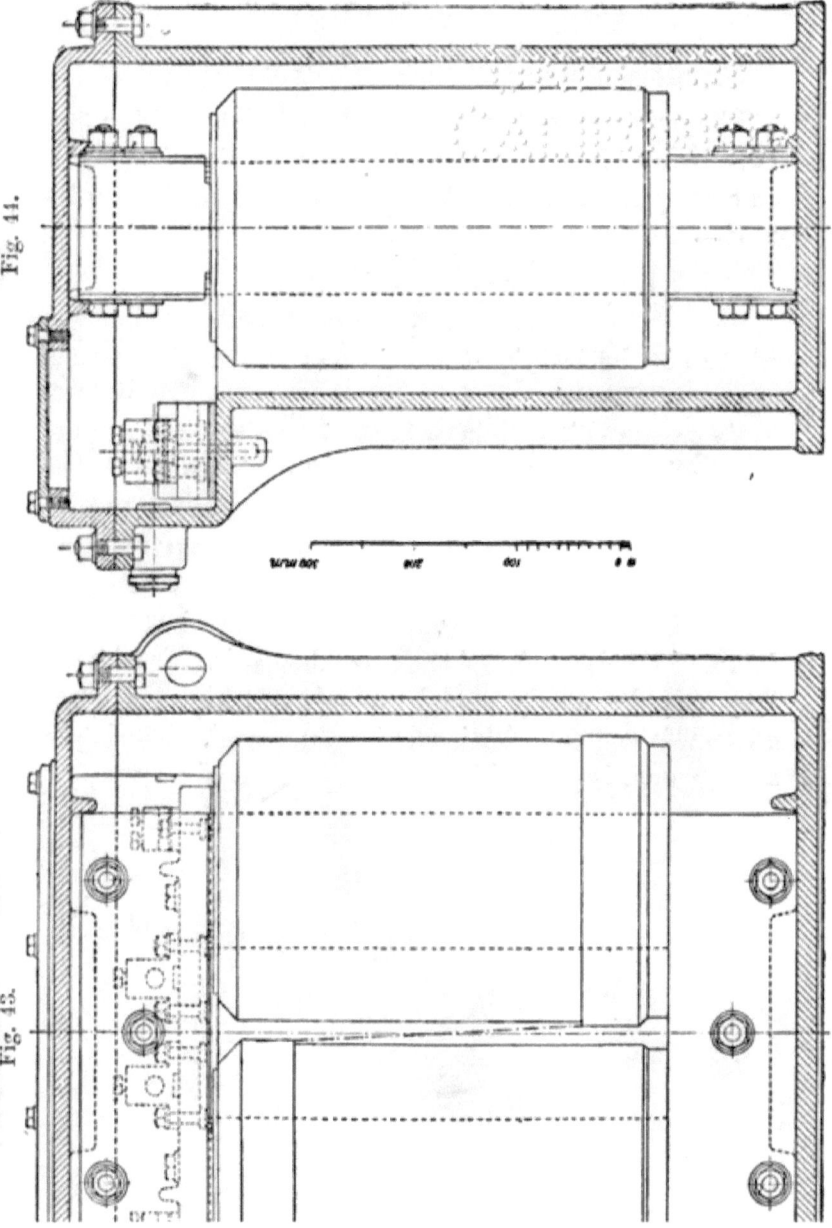

Fig. 11.

Fig. 12.

500 m/m.

300 200 100 0 m

The resistance of the coil must be calculated with reference to its temperature; as a first approximation, based on a temperature of 75° C., the following formula may be used—

$$w = \frac{0\cdot 02\, l}{q} \text{ ohm.,}$$

where l is the length of wire in meters and q its area in square millimetres.

To promote dissipation of heat, the casing may be provided with external ribs or gills. Small internal ribs are also provided to hold the transformer securely. The main cover is fitted with a small auxiliary cover to give access to the terminals without the necessity of breaking the joint of the main cover. The leading-in wires may be taken through stuffing-boxes, as shown in Fig. 41, or they may be simply passed through openings which are afterwards cast out with insulating compound. The latter arrangement is preferable for very large transformers.

CHAPTER VI

Clock diagrams.—The working conditions of a transformer can be represented in a very simple manner by means of so-called vector or clock diagrams. It is thereby convenient to assume for both circuits the same number of turns of wire, which makes the transforming ratio equal to unity. Such an assumption is permissible, as will be seen from the consideration that without altering the gauge of wire on the primary, we can by simply grouping the turns differently bring a sufficient number of wires into parallel connection to produce the effect of a winding of fewer turns, but of stouter wires. If, for instance, the transforming ratio is in reality 2000 to 100, and the high pressure coil has 800 turns, we may assume these 800 turns to be grouped in 20 parallels of 40 turns each. The current will now be 20 times as large as before, and the electro-motive force will be reduced to the 20th part; that is to say, it will not be 2000 volts but only 100 volts. In assuming this alteration in the connections we have not altered in any way the heating, the percentage of idle current, the efficiency, etc., but we have obtained the advantage that the electromotive forces in both circuits are

now of the same order of magnitude, and can therefore be
conveniently represented in our clock diagram to the
same scale. It is obvious that the current must be increased in the same ratio as the electromotive force is
reduced, and that the resistance is reduced in the square of
this ratio.

Working on open circuit.—As an introduction to the use
of clock diagrams we may investigate the simplest case,

Fig. 45.

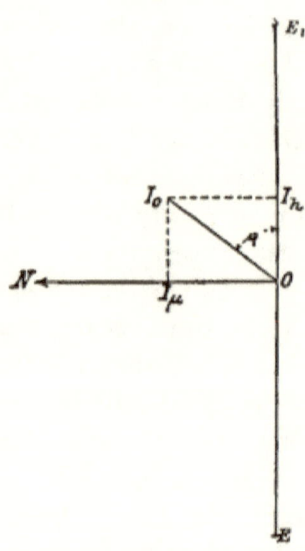

namely, a transformer working on open circuit. Let in
Fig. 45 OI_o represent the idle current to any convenient
scale, and OI_h and OI_μ its two components, which can be
found by calculation as shown in chapter IV. The component I_μ magnetizes the iron of the transformer, the total
magnetic flux being represented to any convenient scale
by the vector ON. The current and flux have of course
the same phase in the diagram. At the moment to which

the diagram refers the projection of the flux on the vertical is zero and the E.M.F. has its maximum value, namely $2\pi \sim Nn\ 10^{-8}$ volt. Since the E.M.F. in the primary must try to prevent the growth of the current (and flux) its direction in the diagram must be vertically downwards. Let this be represented by the line OE plotted to any convenient volt scale. It is obvious that the impressed E.M.F. must be equal and opposite; its vector will therefore occupy the position OE_1. We neglect in the diagram the influence of the resistance of the primary coil since the error thereby committed is insignificant. The diagram then gives OE_1 as the E.M.F. impressed on the primary terminals, OE as the E.M.F. obtained on the secondary terminals, and OI_0 as the primary current, the secondary current being zero.

The power supplied to the primary is $\dfrac{I_h E_1}{2} = \dfrac{I_0 \cos\phi\ E_1}{2}$; or, if instead of using maximum values, we use effective values,

$$i_h\ e_1 = \cos\phi\ i_0\ e_1.$$

The apparent power supplied to the primary is $i_0\ e_1$ and the ratio between effective and apparent power, that is $\cos\phi$, is called the *power factor*.

It is interesting to note that the flux ON which is produced by the current i_0 does neither in position nor magnitude correspond with the flux that would be produced by a constant current of the strength i_0. Yet the equivalent alternating current i_e passes actually through the primary coils, and we should expect that it must produce the corresponding magnetization. This is however not the case. The magnetization is less than corresponds to the ampère turns of exciting power carried by the coil; and it lags behind the current by the angle ϕ.

This apparent contradiction can however be easily explained. The loss on open circuit is due to hysteresis and eddy currents. If we could obtain a magnetically perfect iron, and if it were possible to so design the transformer as to be absolutely free from eddy currents, then i_h = zero and $i_\mu = i_o$. The power factor will also be zero. Assuming

Fig. 46.

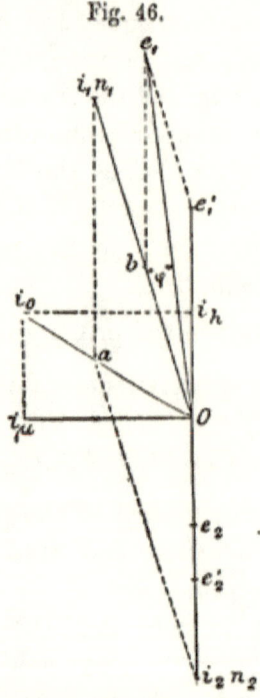

then that it were possible to obtain such a perfect transformer, we could by the addition of a third short circuited winding of the proper resistance so alter it that its working diagram would be the same as that of the transformer practically possible. The only condition to get the two transformers into the same state is that the power which is transformed into heat in the third coil of the perfect transformer equals the power wasted in hysteresis and eddy currents in the imperfect transformer. It is obvious that the current generated in the third coil is on the whole opposed to the primary current, and must therefore weaken the magnetization; and this circumstance explains it— why only one component of the idle current (namely, that which stands at right angles to the component wasting power) is effective in producing magnetic flux.

Working under load.—We may now consider the working diagram of a loaded transformer. For the sake of simplicity we make at first the two following assumptions—first, the transformer has no magnetic leakage; secondly, the load is an absolutely non-inductive resistance. Under these circumstances the secondary current must be proportional to the E.M.F. induced in the secondary coil, and must coincide with it in phase. If i_μ (Fig. 46) represent the magnetizing current, then e'_2 the E.M.F. in the secondary lags by $90°$ behind i_μ and e'_1 the E.M.F. in the primary is in advance of i_μ by $90°$. In this, as in the following diagrams, we assume clockwise rotation of the vectors.

The pressure at the secondary terminals is represented by the line Oe_2; thus the distance $e_2 \, e'_2$ represents the voltage drop due to the resistance of the secondary coil. To find the pressure at the primary terminals as regards magnitude and position in the diagram, we proceed as follows. We plot to an arbitrary scale, the ampère turns in the secondary on the vector of the secondary E.M.F. Let this be the distance $O \, i_2 \, n_2$. To the same scale we plot Oa which represents the ampère turns of the idle current in the primary. It is obvious that Oa must be the resulting ampère turns of the two coils, and since the ampère turns of the secondary coil are known, we find those of the primary coil by drawing the parallelogram as shown in the figure. We thus obtain $i_1 n_1$ and therefore also the primary current i_1. This current is being driven through the apparatus by an electromotive force which must obviously have two components: first, the component Ob required to overcome the ohmic resistance of the primary coil, and secondly, the component Oe'_1 to balance the induced electromotive force. The resultant, also obtained by constructing a parallelogram, is Oe_1, which line therefore represents the electromotive force that has to be supplied at the primary terminals in order that the current i_2 may be drawn from the secondary terminals under an electromotive force e_2. A glance at the diagram shows that $e_1 > e_2$, the difference being in the present case very marked, because for the sake of greater clearness we have exaggerated all losses and assumed too large an exciting power. If the vectors represent effective values, the following relations obtain:

Power supplied equals . . . . . . $e_1 i_1 \cos \phi$

Power given off . . . . $e_2 i_2$

Efficiency η . . . . $\dfrac{e_2 i_2}{e_1 i_1 \cos \phi}$

In good transformers the idle current is only $\frac{1}{20}$ to $\frac{1}{80}$ of the primary current at full load. The distance Oa is therefore, as compared to the distance $O\,i_2\,n_2$, exceedingly small when the transformer is working under full load. It is obvious that the line $O\,i_1\,n_1$, that is the vector of the primary current, becomes nearly vertical, making the angle ϕ so small that its cosine can, without appreciable error, be considered equal to unity. The efficiency therefore becomes $\eta = \dfrac{e_2\,i_2}{e_1\,i_1}$

Magnetic leakage.—Up to the present we have assumed that the transformer is free of magnetic leakage, so that exactly the same magnetic flux passes through the primary and secondary coil. This condition is as a rule not realizable in practice, because the two circuits must be separately wound and insulated from each other, whereby intervening spaces are produced which admit of the passage of leakage lines. The result of this leakage flux is to increase the counter E.M.F. in the driving (primary) coil, and to reduce the E.M.F. in the driven (secondary)

Fig. 47.

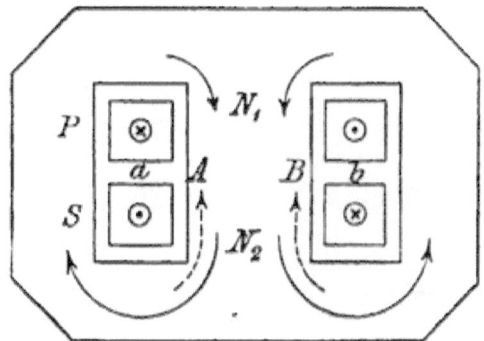

coil, an effect which will obviously be the greater the heavier the transformer is loaded. If the transformer is not loaded at all, that is to say, if it is worked on open circuit, the effect vanishes completely and the ratio between the primary and secondary electro-motive force is equal to the ratio between the primary and secondary number of turns.* At load, however, the transforming ratio rises, by reason of the leakage effect just mentioned, and the secondary voltage drops below its value on open circuit.

---

* A small change in this ratio may occur which is due to electro-static capacity in the high pressure coil, if we transform from the lower to the higher pressure.

Let in Fig. 47 *P* and *S* represent the primary and secondary coils of a shell transformer traversed at any particular moment by the currents indicated by the dots and crosses. The driving coil *P* produces a magnetic field N_1 the lines of which flow in the sense indicated by the arrows shown in full lines. At the same moment the driven coil has the tendency to produce a field the lines of which have the direction shown by the dotted arrows. This field cannot in reality be produced, because the

Fig. 48.

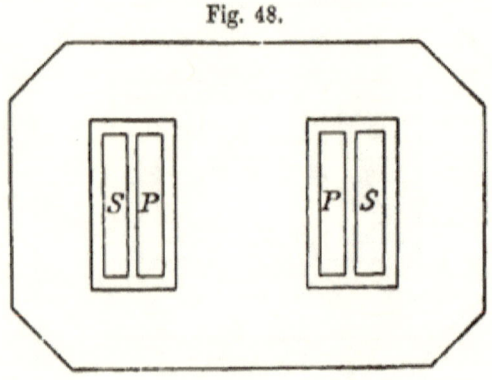

driving power of the primary coil overpowers that of the secondary, so that on the whole the direction of the lines in the core within the coils is vertically downwards, but the effect of the opposition offered by *S*, is to crowd lines of force laterally outward through the sides *AB*, so that the flux N_1 passing through the primary coil is larger than the flux N_2 passing through the secondary coil, the difference N_0 being squeezed laterally through the spaces *ab*. The E.M.F. induced in the secondary coil is therefore not proportional to N_1 but to $N_1 - N_0$ When working on open circuit, the secondary coil carries no current, and there is consequently no tendency to force part of the flux laterally outward through A and B. In this case $N_0 = 0$ and $N_2 = N_1$. The E.M.F. in the secondary is now proportional to N_1, and therefore greater than before, when we assumed the transformer to have been loaded. It follows that with a constant primary pressure, the secondary

Fig. 49.

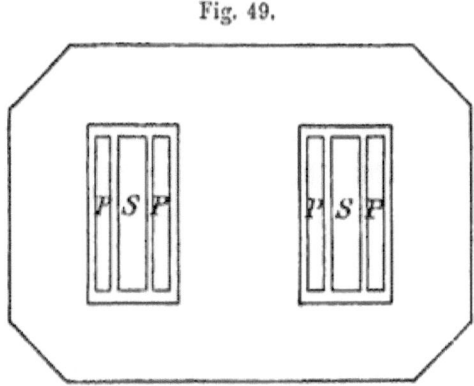

pressure is a maximum on open circuit and decreases as the load on the secondary increases. This so-called

Fig. 50.

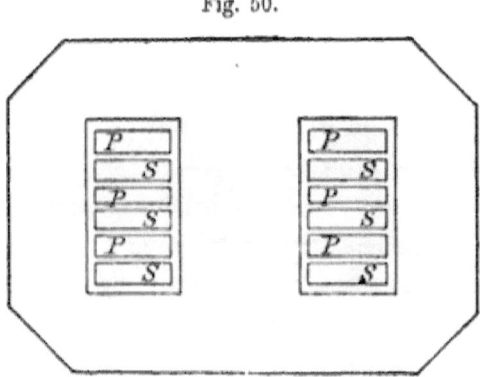

" drop" of a transformer must in most cases be considered as an imperfection, and the question arises as to what means can be used to minimize this imperfection. Since the drop is due to the flux N_0, our aim must be to reduce the latter, and this can be done in either or both of the two following ways. We may reduce the magnitude of the exciting power of each coil, and we may increase the magnetic resistance of the leakage path. Thus the leakage path in Fig. 47 is shorter and wider than in Fig. 48, and we shall in the latter arrangement obtain a smaller drop than in the former. We can carry this improvement a step further by subdividing the primary coil into two parts, which are separated by the secondary coil as in Fig. 49. In doing so we have halved the ampère turns which produce the leakage flux, whilst at the same time still further reducing

Fig. 51.

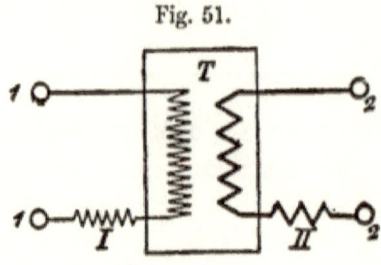

the width of their path. The same object is attained in the arrangement of Fig. 50. Here we have subdivided each coil into a number of parts, each carrying only a corresponding fraction of the total exciting power. The primary coils are sandwiched between the secondary coils, and the leakage flux is produced by a much smaller exciting power, and has a path of much higher resistance than in Fig. 47. This arrangement is also convenient because the coils need not be wound upon each other, but may be wound insulated, and tested independently of each other.

From what has been said above, it will be clear that the effect of leakage can be represented by the addition of two magnetic fields (one interlinked with a coil in the primary and the other with a coil in the secondary circuit) to the main field common to both coils. We may thus imagine the practically possible transformer, which has leakage, replaced by a perfect transformer T, Fig. 51, having no leakage, to which however are added two choking coils I and II, the terminals 1 1, 2 2 being outside of the latter. At full load the choking coil I produces the field N_1 and the choking coil II the field N_2. For the sake of simplicity, we assume the dimensions of the choking coils such that each contains the same number of turns at the corresponding transformer coil. The field common to both transformer coils is N. Then the E.M.F. induced in the primary transformer coil is

$$e_1 = 4{,}44 \sim n_1 \, N \, 10^{-8}$$

and that induced in the choking coil I is

$$e_{s1} = 4{,}44 \sim n_1 \, N_1 \, 10^{-8}.$$

Similarly we have for the secondary circuit

$$e_2 = 4{,}44 \sim n_2 \, N_2 \, 10^{-8}.$$

$$e_{s2} = 4{,}44 \sim n_2 \, N_2 \, 10^{-8}.$$

In plotting these values in a clock diagram we must not forget to place e_{s2} at right angles to i_2 and e_{s1} at right angles to i_1.

Working diagram of transformer having leakage.—We are now in a position to draw the clock diagram of a transformer having leakage. Let for the present its load be now inductive (glow-lamps for instance), and assume that we have, as mentioned in the beginning of this chapter, arranged the primary winding so as to have the same number of turns as on the secondary coil.

Let $O i_2$ in Fig. 52 represent the secondary current, $O e_{k2}$

I

the pressure at the secondary terminals, e_{k2} e'_2 the loss of pressure due to the ohmic resistance of the secondary, then Oe'_2 must be the resultant of the E.M.F. induced by the main field N and the E.M.F. of self-induction due to the leakage field N_2. The vector of the latter must lie in such

Fig. 52.

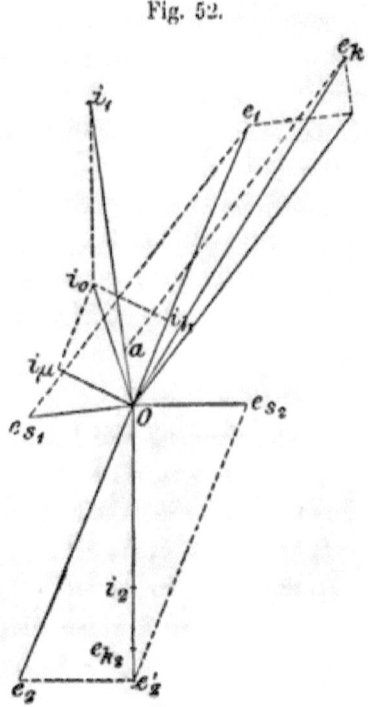

position that e_{s2} tries to prevent the decrease of i_2; that is to say, it must be drawn horizontally to the right as shown at e_{s2} in the diagram, which is plotted to the same volt-scale as Oe_{k2}. By constructing the parallelogram we obtain Oe_2 the E.M.F. which must be induced in the secondary coil in order that the current i_2 may flow under the terminal pressure e_{k2}.

The magnetizing current i_μ must be at right angles to e_2 and in advance of it, whilst the current representing loss of power i_h must be in line with $e_2 e_1$ and of the same sense as e_1. The position and magnitude of the idle or no-load current is thus completely defined. We obtain for it in the diagram the vector i_o. The vector of the primary current we find by combining i_g and i_2. This gives i_1. The E.M.F. of self-induction (due to leakage) in the primary must stand at right angles to i_1 and must follow it. In our diagram, e_{s1} must therefore be drawn to the left. The E.M.F. supplied to the primary terminals must obviously contain three components.

One component must be equal and opposite Oe_2. This is given by the vector Oc_1.

One component must be equal and opposite Oe_{s1}, and one component must be provided to overcome the ohmic resistance. Let the vector of the latter be Oa.

By adding these three components graphically we obtain the point e_{k1}. Oe_{k1} is the vector of the E.M.F. supplied to the primary terminals. A glance at the diagram shows that e_{k1} is greater than e_{k2}, the difference being the more marked the greater are the ohmic resistances of and the E.M. Forces of self-induction in the two windings. In both respects the diagram Fig. 52 has been exaggerated, so that the influence of each part may be more clearly seen.

It is interesting to investigate the case of a transformer the secondary terminals of which are short-circuited by a stout copper wire and ampèremeter, thereby making $e_{k2} = O$. We assume the primary E.M.F. to be so adjusted that this ampèremeter shows the normal secondary current corresponding to full load under normal working conditions. The diagram then assumes the form shown in Fig. 53. The lettering is the same as in Fig. 52. It will be seen from this diagram that although no pressure is obtained at the secondary terminals, a pressure equal to e_{k1} must be supplied to the primary terminals in order that the current i_2 may flow through the short circuit.

If, as is always the case in modern transformers of good

Fig. 53.

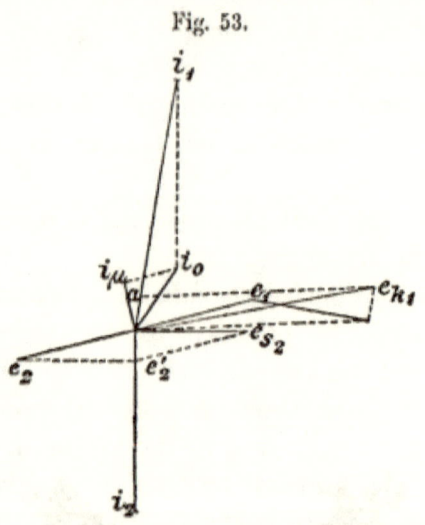

design, the resistance of the windings is very small, then the vector of e_2 is nearly horizontal and that of i_1 is nearly vertical. The points e_1 and e_{k1} have then also approximately horizontal vectors, and we have very nearly

$$e_{k1} = e_{s2} + e_{s1}.$$

With a symmetrical arrangement between the two windings (and the assumption that the number of turns is the same in both) we have $e_{s2} = e_{s1}$ and

$$e_{k1} = 2\, e_{s2}.$$

The E.M.F. of self-induction can thus be found by a very simple experiment. We short-circuit the secondary

terminals by means of an ampèremeter (having itself as
little induction as possible), and supply the primary
terminals with current of normal frequency and such
E.M.F. that the normal secondary current is indicated on
the ampèremeter. One half the E.M.F. supplied to the
primary equals the E.M.F. of self-induction in the primary
winding at normal load. The E.M.F. of self-induction in
the secondary winding is equal to this value divided by
the transforming ratio. Take as an example the case
of a 10 Kwt. transformer wound for a ratio of 2000
volts to 100 volts. In testing this transformer, as above
explained, it is found that 400 volts must be supplied on
the primary at $\sim$ = 50 in order that 100 ampère may be
driven through the short-circuit. We have then e_{s1} = 200
and e_{s2} = 10 volts.

The experiment may also be used to determine the
co-efficient of self-induction of each coil.

Let L_2 be the co-efficient for the secondary and L_1 that
for the primary coil, then

$$e_{s2} = 2\,\pi \sim i_2\,L_2$$
$$10 = 6{\cdot}28 \times 50 \times 100 \times L_2$$
$$L_2 = 3{\cdot}18 \times 10^{-4}\ \text{Henry}$$

For the primary coil e_{s1} = 200 and i_1 = 5

$$L_1 = 400\,L_2$$
$$L_1 = 0{\cdot}127\ \text{Henry}$$

It must be noted that these values refer only to the
transformer with short-circuited secondary.

Voltage drop.—If by making the experiment above
described we have found how much E.M.F. is produced
by magnetic leakage in each coil, we can use this informa-
tion to determine the voltage drop at various loads. In
this determination it is convenient and permissible to
assume exact opposition in the phases of primary and

secondary current. Modern transformers with closed magnetic circuit require so little magnetizing current, that even at moderate loads this assumption is very nearly true. Let in Fig. 54, OA represent the pressure at the secondary terminals, AB the ohmic loss of pressure, $BC = c_{s_2}$ the E.M.F. due to self-induction; and therefore $OC = c_2$ the E.M.F. induced in the secondary. Let

Fig. 54.

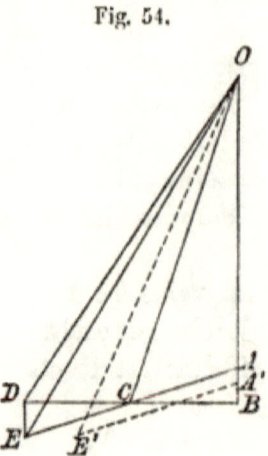

the transforming ratio be reduced to unity, then $OC = c_1$ is also the E.M.F. induced in the primary, and with symmetrical windings $CD = BC$ the E.M.F. self-induction in the primary, so that $c_{s_1} = c_{s_2}$. The ohmic voltage loss in the primary is $DE = AB$ if the losses are equally divided between the two windings as required by a good design. The line joining A, C and E is therefore a straight line, and its inclination to the vector of secondary terminal pressure is the same for all loads. At a smaller load, for instance, producing the ohmic loss $A'B$, the terminal pressure would be OA' in the secondary and OE' in the primary. The ratio in the length of the lines AE and $A'E'$ is the same

as that in the lines *AB* and *A'B,* and the length of the line
A E is directly proportional to the load.

Let us now assume that we are able to vary the primary
E.M.F. in any way which may be required to keep the
pressure at the secondary terminals constant for all loads.
We draw the line *A E* (Fig. 55) for full load, and make
an ampère scale which corresponds with this length at full
load, then we can, by using this scale, mark off on the line

Fig. 55.

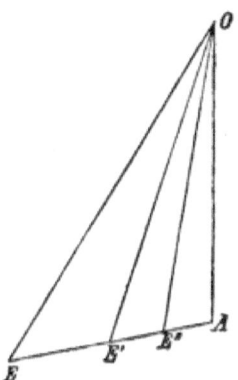

A E the points *E'*, *E''*, etc., corresponding to other loads, and
thus find the primary E.M.F. vector *OE'*, *OE''*, etc., corresponding to these loads. It is thus possible to determine
the primary E.M.F. as a function of the load, if the secondary terminal pressure is to be a constant.

This is however not the case generally met with in
practice. As a rule the E.M.F. in the primary or supply-
circuit is constant, and it is required to find the secondary
terminal pressure at each load. This problem can also be
solved graphically in a very simple manner.

Graphic determination of drop.—It has already been
shown that in all the triangles *OAE*, *OA'E'*, etc., the

obtuse angle at A, A', etc., is the same.   The longest side
of the triangle represents the E.M.F. impressed on the
primary, and the shortest side the load, that is the current
in the secondary.   We may now imagine all the triangles
in Fig. 55 so enlarged or reduced that all the points E lie
on a circle described round O as centre, with a radius equal
to the impressed E.M.F.   Let OE in Fig. 56 represent
this E.M.F. at full load (current represented to a suitable
scale by the lengths AE) and E', E'' the positions of E for

Fig. 56.

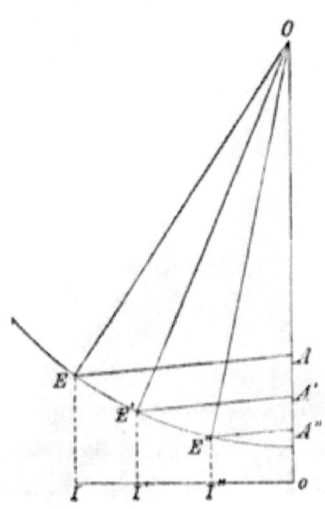

smaller loads, then the length OA, OA', OA'', etc., gives the
corresponding pressures at the secondary terminals.   As a
matter of convenience we may also plot the secondary current on a horizontal oI to a suitable scale, and find the
points E by projection from the points I, as shown by dotted
lines.

If we apply this method to our previous example of a
10 Kwt. transformer, and assume that at full load the ohmic

loss in each winding is 1 per cent., and the E.M.F. of self-induction 10 per cent., we find that the line OE has an inclination to the horizontal of 1 : 10. If at full load the secondary terminal pressure is to be 100 volts, then OA must be 100 on the volt scale, and the horizontal distance between A and E, 20 on the same scale. The ampère-load is to be plotted on the horizontal oI by using a scale on which the length oI represents 100 ampère. The current is plotted on oI to the same scale, and by projecting the corresponding points, first to the circle and then to the vertical parallel to EA, we find the terminal volts OA', OA'', etc. This construction, carried out for various loads, gives the following results, the impressed E.M.F. being constant.

Ampères in secondary	0	25	50	75	100	200
Terminal pressure	103·8	103·2	102·35	101·3	100	92

The drop between no load and full load is thus 3·8 volts. The drop between full load and 100 per cent. over load (which the transformer is perfectly able to stand for a short time) is 8 volts more, or a total between no load and double the normal full load of 11·8 volts.

Up to the present we have assumed that the load is non-inductive. It remains yet to extend the investigation to cases in which the secondary circuit has also self-induction, or capacity, or both. Self-induction is introduced if the secondary current is used for feeding motors or arc lamps, in which cases there is developed an E.M.F. at right angles to the current. The pressure at the secondary terminals must therefore have a component equal and opposite to this E.M.F. of self-induction, and this component must be in advance over the current by 90°. Let in Fig. 57 OA represent the secondary current, OB the power component of the secondary pressure, and OC the

counter **E.M.F.** produced by self-induction. The secondary
pressure is then represented by the vector *OD*, which ad-
vances over the current by the angle ϕ. We call *cos* ϕ the
power factor of the motors or arc lamps worked by the

Fig. 57.

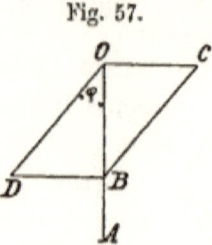

secondary current, and *BD* is the wattless component of
the secondary pressure.

It is conceivable that the secondary circuit contains
partly appliances having only resistance, and partly other
appliances having resistance and self-induction. This is
the case if to the same leads are joined glow-lamps and
arc-lamps or motors. In a distributing system of 100

Fig. 58.

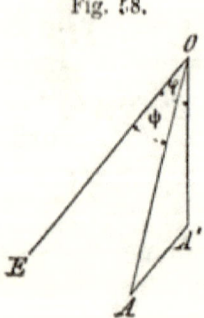

volts the glow-lamps would be arranged in simple parallel
and the arc-lamps in parallel series of two or three lamps.
In Fig. 58 *OE* is the vector of secondary pressure, ϕ the
lag of that part of the circuit which has self-induction, and

OA' the corresponding current. The current taken by the glow-lamps is in phase with the pressure, and must be represented by a vector which is parallel to OE. Let this vector be $A'A$, then OA is the total current supplied to the circuit and ψ its lag behind the pressure. A glance at the diagram shows that $OA \prec OA' + A'A$ and that $\psi \prec \phi$. If then we measure separately the current supplied to the glow-lamps and to the arc-lamps, and also the current in the undivided part of the circuit, we shall find that the latter is less than the sum of the other two currents. Let for instance the power factor of the arc-light circuit be 0·71 ($\phi = 45°$), and let there be 5 parallel series of lamps, each taking 15 ampère, then $OA' = 75$ ampère. In addition to the arc-lamps let us insert a number of glow-lamps, taking collectively 32 ampère. The total current is then not 107, but only 100 ampère, as will be found on drawing Fig. 58 to scale. The power factor of the whole circuit is then $cos\ \psi = 0·85$. The transformer is thus apparently loaded to 10 Kwt.; in reality, however, only to 8·5 Kwt. The reduction in the load is brought about by the phase difference between pressure and current. We have now to investigate what effect this phase difference has on the terminal pressure or on the ratio of primary and secondary terminal pressure.

Let in Fig. 59, OA represent the secondary current and OB the terminal pressure. The E.M.F. induced in the secondary winding must obviously contain the following three components. First, OB; secondly, BB', to cover the ohmic loss; and thirdly, $B'C$, to counteract the E.M.F. of self-induction. The vector of the secondary E.M.F. is therefore OC, and by our assumption of an equal number of turns in both coils, OC is also the E.M.F. induced in the primary coil. The E.M.F. impressed on the primary terminals must also have three components; namely, OC, CD, to counteract self-induction, and DE, to cover the ohmic loss. We thus obtain the vector of the E.M.F. impressed on the primary terminals OE. The inclination of the straight line BE is, as before, determined by the ratio between resistance and reactance.

Fig. 59.

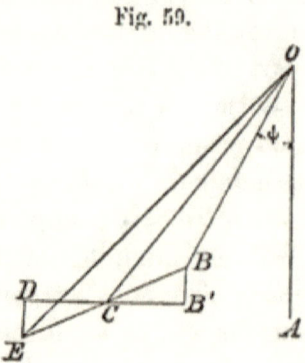

With a mixed load of glow-lamps and arcs the angle ψ varies with the number of lamps of each kind in use at any time. If however the load consists only of arc-lamps which are switched in or out in series of two or three, then the lag (which we will call ϕ in this case, compare Fig. 57) remains the same for all loads. The current may, as before, be plotted on a line oI drawn parallel to BE (Fig. 60), the scale being so chosen that oI represents full load. The pressure on the secondary terminals is for full load OB. For a smaller load oI_1 the construction shown in the diagram gives a greater pressure, namely OB'. By comparing this diagram with Fig. 56 it will be seen that the drop is greater when the load is inductive. This may also more precisely be seen if we extend the example previously given to a case where the load consists of arc-lamps. We investigated the behaviour of a 10 Kwt. transformer, in which the ratio of reactance to resistance was 10 : 1. Let us now connect the same transformer to an arc-light circuit in which $cos \phi = 0.71$ ($\phi = 45°$), and construct the diagram Fig. 60 for various loads. This gives the figures contained in the following table. To facilitate the comparison the figures of the previous table are repeated.

Secondary current in ampère		0	25	50	75	100	200
Pressure at secondary	$cos \phi =$	1	103·8	103·2	102·35	101·3	100	92
terminals	$cos \phi =$	0·71	103·8	99·5	95·2	91	86·8	68

This transformer (2 % resistance and 20 % reactance) has on an inductionless resistance a full load drop of less

Fig. 60.

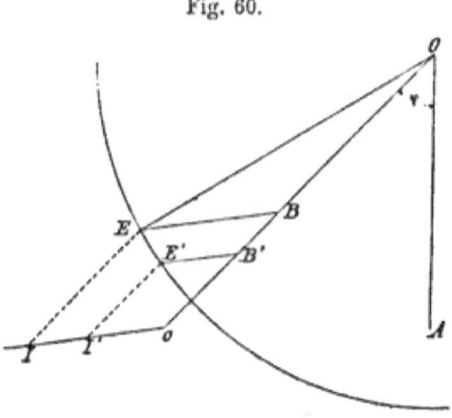

than 4 per cent. Although such a performance is not quite as good as might be desired, it is tolerable; for an inductive load this transformer is however quite unfit, since the drop amounts to as much as 17 per cent. To render this transformer suitable for motor work, the geometric arrangement of the winding would have to be so altered as to considerably reduce magnetic leakage. If tested on short-circuit, the full secondary current must be obtained with a primary E.M.F., which at the outside must not exceed 10 per cent. of the normal supply voltage. Thus the drop at full load with $\phi = 45°$ would be about 9 per cent., which is tolerable for motor-work, but already too high for arc-lighting. Transformers for arc-lighting, or a joint service of lighting and power purposes, must be so designed that the product of current and reactance does not exceed 5 per cent. of the normal supply voltage. This result can be attained by proper grouping and thorough subdivision of the coils.

Our investigation thus far has shown that with a constant E.M.F. impressed on the primary terminals, the pressure on the secondary terminals drops as the load increases, the drop being due to the ohmic resistance in both coils, and to their reactance. We have also seen that, other things being equal, the drop is the greater, the smaller the power factor of the circuit to which the transformer supplies current.

The case where the circuit has capacity must now claim our attention. Capacity also reduces the power factor, and it may perhaps be thought that also in this case the reduction in the power factor would be accompanied by an increase of voltage drop. This, however, is not the case. On the contrary, the introduction of capacity into the circuit diminishes the drop, and may under certain circumstances even convert it into a rise of pressure. To facilitate the investigation, we assume at first that the apparatus supplied with current by the transformer has only resistance and capacity, but no reactance. Let the capacity be a shunt to the resistance (a concentric cable feeding glow-lamps is a practical illustration of this case). If E is the maximum E.M.F., K the capacity in Farad, then EK coulombs are charged into and discharged from the condenser twice every period, the charges being alternately positive and negative. Let us consider the moment when the E.M.F. has attained its positive maximum value, and commences to decrease. The condenser is then completely charged by the positive current which has up to that time been flowing into it. As soon as the E.M.F. begins to decrease, the period of discharge begins, the current being now negative, although the E.M.F. still remains positive for a quarter-period longer. The negative current attains its maximum value at the moment when the E.M.F. passes through zero. It is therefore in advance over the E.M.F. by a quarter-period.

Fig. 61.

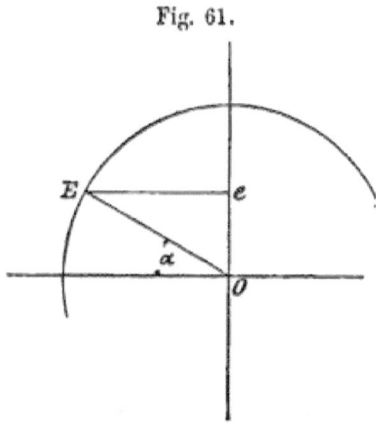

In Fig. 61 *OE* represents the position of the E.M.F. vector at time t corresponding to the angle a, and e the instantaneous value of the E.M.F. After the lapse of an infinitely short time dt the instantaneous value has increased by $de = \dfrac{d}{dt} E \sin a \, dt$, and the charge of the condenser has been increased by the amount $i \, dt$, i being the current at time t which has flown under the potential difference de. We have therefore

$$i\,dt = K\,dc$$

$$i = K\frac{dc}{dt}$$

The differential quotient $\frac{dc}{dt}$ is determinable if the shape of the E.M.F. curve is known. Let this be a sine function, then

$$\frac{de}{dt} = E \cos a \frac{da}{dt},$$

and since $da = 2\pi \sim dt$ we have

$$\frac{dc}{dt} = E\,2\pi \sim \cos a$$

$$i = KE\,2\pi \sim \cos a.$$

The condenser current attains its maximum value at all values of a for which $\cos a = \pm 1$; that is to say, for $a = 0$, $a = \pi$, etc. It is zero for $a = \frac{\pi}{2}$, $a = \frac{3}{2}\pi$, etc. Since the opposite is the case with regard to the instantaneous value of the E.M.F., we see that the vector of the condenser current must stand at right angles to the E.M.F. vector, and be in advance over it by 90°. The maximum value of the condenser current is

$$I = KE\,2\pi \sim,$$

and its effective value

$$i = \frac{KE}{\sqrt{2}}\,2\pi \sim$$

Lines $\frac{E}{\sqrt{2}}$ is the effective value of the E.M.F., which we may designate by e; we have also

$$i = Kc\,2\pi \sim$$

In this formula i is given in Ampère, K in Farad, and e in Volt. The usual unit of capacity is, however, not the Farad, but the Microfarad (a million times smaller), and by adopting this unit we have

$$i_k = K e\, 2\pi \sim 10^{-6},$$

the index $_k$ being added to the sign for the current, to show that the latter is a condenser current, which, leading over the E.M.F. by 90°, carries no power. Let i_w be the watt or power component of the current, the phase of which coincides with the phase of the E.M.F., then

$$i_w = \frac{e}{W}$$

W being the ohmic resistance of the apparatus to which current is supplied. By reference to Fig. 62 it will be seen that the total current i is given by the expression

$$i = \sqrt{i_k^2 + i_w^2}$$

Fig. 62.

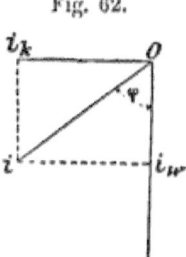

Oe is the E.M.F. vector, Oi_k the vector of the condenser current, and Oe_w the vector of the power current. It is important to note that the diagram only represents the case where capacity and resistance are in parallel, and the condenser circuit has so little resistance that it may be neglected. This is approximately the case of a concentric cable connected at one end to the transformer and at the other to glow-lamps. The two conductors form the inner and outer coating of the condenser, which is being charged and discharged by the condenser current i_k. If all lamps are switched out $W = \infty$ and $i_w = 0$. In this case $\phi = 90°$ and $i = i_k$. The cable takes only the condenser

K

current. As lamps are being switched on i_{μ} increases, ϕ decreases, i increases, and the power factor increases also.

If arcs instead of glow-lamps are connected to the far end of the cable, there will be reactance in addition to resistance. The E.M.F. of self-induction is

$$e_s = 2\,\pi \sim L\,i_{\mu}$$

and its phase lags by 90° behind the phase of the current. The corresponding component of the E.M.F. impressed on the circuit must therefore have this value, but be 90° in advance of the current. The watt component e_{w} is of

Fig. 63.

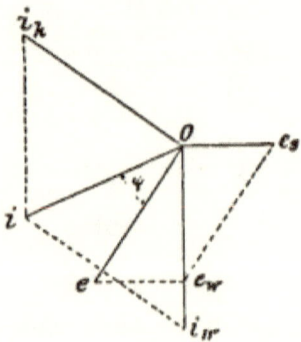

course co-phasal with the current. In Fig. 63 i_{μ} represents the current flowing through the inductive resistance, e_{w} the watt component of the E.M.F., and e_s the E.M.F. of self-induction. By a parallelogram we find Oe, the vector of the E.M.F. This not only drives the current through the inductive resistance, but it also produces the condenser current i_k leading by 90°. The total current supplied by the transformer is the resultant of i_{μ} and i_k and its vector is Oi. Accordingly as reactance or capacity preponderates, the current will lag behind or lead before the E.M.F. In the diagram the relations are so chosen that the current leads.

We have up to the present assumed that the condenser forms a shunt to the apparatus absorbing power. This is indeed the most frequent case, but it is also possible that the circuit is interrupted by a condenser, which is then in series with the apparatus absorbing power. A case of this kind occurs if a liquid resistance is used in testing transformers. A barrel containing salt water or acidulated water, and two sheets of lead as electrodes, or an iron trough containing an alkaline solution, and iron plates for electrodes make very efficient resistances, capable of taking up large amounts of power. Liquid resistances are for these reasons very often used in lieu of solid resistances. It is well known that a metal plate dipping into a liquid acts as a condenser of very large capacity, and in using a liquid resistance we introduce therefore into the circuit a capacity in series with the resistance.

Fig. 64.

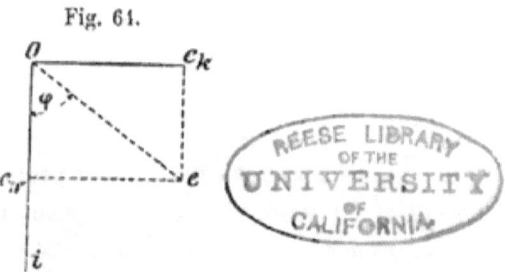

Let in Fig. 64 Oi represent the current passing through the liquid resistance, and Oc_w the watt component of the E.M.F. The E.M.F. required to produce the condenser current i is $c_k = i : K 2 \pi \sim$ in absolute measure. It lags behind the current by 90°, as shown in the diagram. The resultant E.M.F., that is the E.M.F. which must be impressed on the terminals of the liquid resistance in order that the current *i* may flow through it is *Oc*. It is obvious that also in this case the current leads over the E.M.F. by a certain angle (in the diagram the angle ϕ), and that the power factor of the liquid resistance is smaller than unity.

In the foregoing it has been shown how the phase-difference between current and pressure can be determined for every load if the electrical constants of the apparatus receiving power are known. In this manner the working conditions of the transformer can in every case be determined.

Fig. 65.

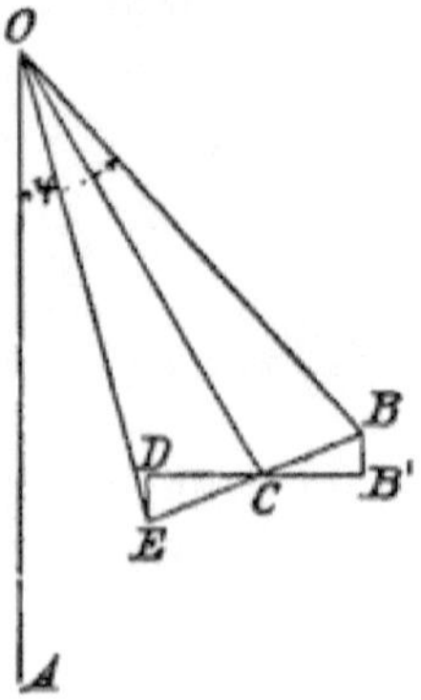

Those cases where the terminal pressure leads as compared with the current have already been investigated, and we have now to extend the investigation to those cases where the current leads over the pressure at the secondary terminals of the transformer, that is to say, where the phase angle ϕ is negative. The construction previously given can obviously also be applied now. In Fig. 65 the current is given by the vector *OA*, and the terminal pressure by the vector *OB*. The ohmic drop in the secondary is *BB'*, and must obviously be parallel with *OA*. The E.M.F. of

self-induction in the secondary is $B'C$, which is at right
angles to OA. The E.M.F. of self-induction in the
primary is CD, and the ohmic drop in the primary is
DE. We thus obtain OE as the vector of the E.M.F.,
which must be impressed on the primary terminals of the
transformer. Since BB', DE, and $B'D$ are all proportional to the current, the inclination of the line BE
remains the same for all loads, and its length may to a
suitable scale be made to represent the secondary current.

Fig. 66.

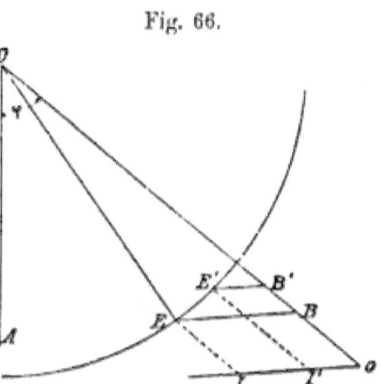

We assume for the present that the power factor is
independent of the current. In this case the pressure at
the secondary terminals may be found for different currents
by the construction shown in Fig. 66, in which OA
represents the direction of the current vector, and Oo
that of the secondary terminal pressure. The inclination
of Io, on which we measure the secondary current, is
fixed by the electrical constants of the transformer. Io
must be parallel to EB in Fig. 65. From O as centre we
describe a circle with a radius equal to the E.M.F.
impressed on the primary terminals (the transforming
ratio is supposed to be unity). To find the position of

the primary E.M.F. vector we measure from o to the left a distance representing the secondary current, and find thus the point I. From I we draw a parallel to oO, and find E. Drawing EB parallel to Io we find B. Then OB is the pressure at the secondary terminals. For a smaller current I' the same construction gives the pressure OB'.

As will be seen from the diagram, the terminal pressure increases with the current. We have therefore with increasing load, not a drop, but a rise of terminal pressure. The rise of pressure depends, amongst other things, on the angle of lead ϕ. If this is smaller the rise will also be less marked, as will easily be seen from the diagram, and it is obvious that there must be a certain phase angle, for which there is neither rise nor drop of pressure, but exactly the same pressure at full load and at no load. For smaller values of ϕ there will be a drop, but this will still be less marked than would be the case if the current lagged behind the E.M.F. We thus come to the conclusion that the drop depends, not only on the magnitude, but also on the nature of the load. It is greatest for inductive loads, smaller for non-inductive loads, and smallest (or even negative, *i.e.* a rise) on loads having capacity. For this reason the use of a liquid resistance in testing transformers for drop is misleading. The drop so measured is always smaller than that which would be found on a lamp-circuit, and it may even happen that the drop on a liquid resistance is negative, that is to say, that the pressure rises when the load is increased. This will happen if the transformer has much magnetic leakage. Thus the worse the transformer the more favourable will it appear when tested on a liquid resistance.

Up to the present we have assumed that the angle of

lead or lag ϕ is the same for all loads, and determined the pressure at the secondary terminals as a function of the secondary current. In most cases it is, however, only of importance to know the drop or rise at full current, since the fitness of a transformer for practical use must naturally be judged by its behaviour at full, and not at some intermediate, load. On the other hand, it is important to know how the secondary pressure depends on the power factor of the apparatus to which the transformer supplies current. The same transformer may be intended for use in connection with apparatus of different power factor, and it will depend on the change in terminal pressure between full load and no load whether such use is practically possible.

The problem we have to solve is therefore the following— Given a transformer of known resistance and reactance. The primary or supply E.M.F. is constant. Find the pressure at the secondary terminals with full secondary current, but for different values of phase angle ϕ. The graphical solution of this problem is extremely simple, and follows as a matter of course from what has been said in connection with Figs. 59 and 65. For a constant ampère-load the length of the line $E\,B$ is constant, the only variable in the diagram being the angle ϕ which the vector of terminal pressure forms with the current vector. The inclination of $B\,E$ is also constant. As ϕ changes E takes different positions on the circle of primary E.M.F., and the locus of B must therefore also be a circle of the same radius, the centre of which has relatively to O the same displacement as B has to E.

Let in Fig. 67 the vertical represent the current vector, $O\,S$ the E.M.F. of self-induction at full current, and $S\,o$ the ohmic loss at full current in both windings; then $O\,o$

is equal and parallel with *E B* of Fig. 60, and *o* is the centre of the second circle just mentioned. For a positive phase difference (current lagging behind E.M.F.) the secondary terminal pressure *O B* is smaller than *O E*, its value at no load. For a negative phase difference ϕ_1 (current leading before the E.M.F.) the secondary terminal pressure $O B_1$ is greater than its value at no load. With a certain negative phase difference ϕ_2 the terminal pressure at full load is exactly the same as at no load, and we have neither a drop nor a rise of voltage. This phase angle is given by the point of intersection B_2 of

Fig. 67.

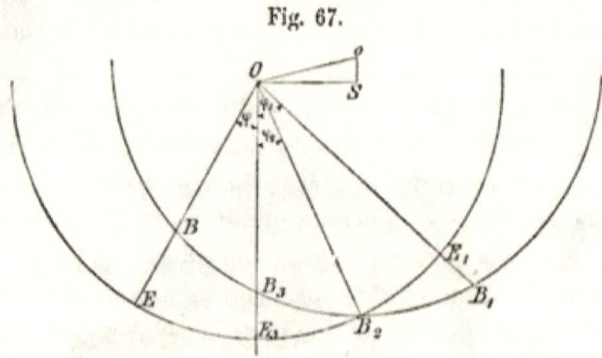

the two circles. The drop or rise can be directly measured in the diagram; it is the length included between the two circles. If the load is an inductionless resistance having no capacity $\phi = o$ and the drop is given by the distance E_3 to B_3. The diagram shows very clearly how the drop increases as the lag of current behind E.M.F. becomes greater in consequence of the greater reactance in the apparatus to which the transformer supplies current.

The length of the line *O o* is, as already pointed out, a measure for the secondary current. If the load is reduced o is shifted accordingly, and the same construction can be made for every load. We can thus determine the secondary terminal pressure for every working condition of the transformer. Such a determination is to be preferred to a direct measurement of the drop, for the following reasons. In the first place, to make the direct measurement reliable, the apparatus serving as a load to the transformer must have exactly the same power factor as obtains in the apparatus to which the transformer will have to supply current. A liquid resistance can therefore, as a rule, not be used in the test. Solid resistances and lamps are not only more cumbersome, but their adjustment to produce one particular lag requires the use of choking-coils or similar appliances, and this complicates the test. In the second place, the direct testing of large transformers requires an amount of power which is not always available; and lastly, if these difficulties are overcome, we have to face the further difficulty of having to determine a comparatively small difference between two readings of large absolute value, which cannot be done with great accuracy. For these reasons it is better not to attempt any direct determination of the drop, but to solve the problem indirectly, in the manner described with reference to Fig. 67.* As an example to show the application of the indirect method, we take a 60 Kwt. transformer having a ratio of 3000 to 200 volt. Resistance of primary 0·9 ohm; ohmic loss 18 volt. Resistance of secondary 0·0036 ohm; ohmic loss 1·08 volt. By reducing the transforming ratio to unity we would have an ohmic loss in the primary

of $18 . \dfrac{200}{3000} = 1\cdot20$ volt. To determine the length $O o$

---

* See the author's paper on this subject (*Elektrotechnische Zeitschrift*, 1895, p. 260), and discussion thereon.

in Fig. 67 we have therefore the following data:—

Ohmic loss in primary	1·20
„ „ „ secondary.	1·08
„ „ „ total	2·28

We short-circuit the secondary through an ampère-meter, and supply as much primary E.M.F. as will just suffice to produce the full secondary current of 300 ampère. This experiment shows that 255 volt must be supplied to the primary terminals. This corresponds to 17 volt at the transforming ratio of 1 : 1. We have therefore in Fig. 67 $O\,o = 17$ and $S\,o = 2·28$, whilst $O\,E$ is 200. We can now design the diagram, Fig. 68. $O\,A$ is the current vector. On this we mark off the power factor cos ϕ. The corresponding position of the vector of E.M.F. is $O\,E$, and the terminal pressure which we scale off on $O\,E$ is 187 volt. In a similar manner we determine the terminal pressure for all other values of cos ϕ. The result is given in the following table.

60 Kwt.-transformer 3000 : 200 volt on open circuit.

Pressure at secondary terminals with 300 ampère in secondary and power factors varying from 100 to 50 per cent.

Power factor in per cent.	100	99	90	80	70	60	50
With leading current . .	197	200	205	207	210	212	213
With lagging current . .	197	195	190	188	187	186	185

If used on a glow-lamp circuit this transformer would at full load have a drop of only $1\frac{1}{2}$ per cent.; if used on a circuit containing arc lamps or motors the power factor of which is about 0·70 to 0·80, the drop would be approximately 6 per cent.

The diagram Fig. 68 leads to some interesting deductions. In the majority of cases the circuit has not capacity, but reactance, and the following remarks apply to these cases, that is to say, to the left-hand side of the diagram. If we could build a transformer which has absolutely no magnetic leakage, then OS would be zero, and o would lie vertically above O. The inner circle would then approach the outer circle more closely as we go to the left. In other words, the drop would be greatest for an inductionless, and smaller for an inductive resistance. This case is, however, unattainable in practice,

Fig. 68.

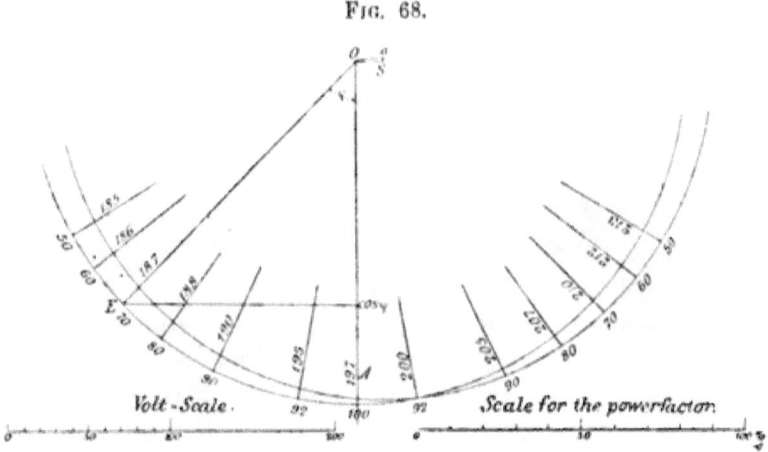

for we can never reduce magnetic leakage to zero. The reactance produced by magnetic leakage can, however, with a very careful design, be made very small, especially for low periodicities. Imagine that we have reduced the reactance so far as to be equal to the resistance, then $OS = So$, and Oo includes with OA an angle of $45°$. The distance between the two circles would then be approximately the same for all values of ϕ. We should thus obtain a transformer which has approximately the same drop for all values of the power factor.

As a rule, the reactance is, however, considerably greater

than the resistance, and the two circles diverge towards the left. As a consequence the drop increases as the power factor decreases. If the same transformer is used for a high and a low frequency, the pressure at the secondary terminals will at full current be lower in the former case. The E.M.F. of self-induction is for both windings,

$$O\,S = 2 \times 2\,\pi \sim L_2\,i_2,$$

that is to say, $O\,S$ is proportional to $\sim$. The higher the frequency, the greater is the divergence between the two circles. It must also be borne in mind that the power factor of the apparatus to which the transformer supplies current (motors or arc-lamps) is lower at the higher frequency, and in consequence the vector of E.M.F. in our diagram is shifted the more to the left, the higher the frequency. Both causes conspire to increase the drop. If then the transformer is intended to feed not only glow-lamps, but also motors and arc-lamps, the frequency should be chosen as small as compatible with the proper working of alternating current arcs (45 to 50 complete cycles per second). This frequency is also advisable on account of certain reasons connected with the design of non-synchronous motors.

CHAPTER VII

The dynamometer.—We have up to the present always supposed that current and E.M.F. follow a sine law. This assumption, although very convenient for the analytical or graphic treatment of alternate current problems, is rarely in accordance with fact, and the question therefore arises what errors are introduced by assuming that current or E.M.F. change according to a curve which is different from that actually obtaining. This question will be answered by investigating whether an ampèremeter, if applied to the measurement of a current of irregular curve, gives the true effective value of the current, and a watt-meter gives its true power. To investigate this matter, it is in the first place necessary that we come to an understanding as to what is exactly meant by the term " effective current." Let us assume that we have two glow-lamps of exactly the same construction, and that one of these is fed by a continuous current, and the other by an alternating current. If both lamps are brought to the same state of incandescence, and absorb therefore an equal amount of power, then the effective strength of the alternating current will obviously be equal to that of the continuous

current. Since both lamps must have the same temperature, the resistances must also be equal. Let this be W,
then the work done in the time T is $I_0^2 W T$, if I_0 is the
strength of the continuous current. The work done on
the other lamp is obviously $\int_0^T I^2 W \, dt$, if I denotes the
momentary value of the current, which is, of course, a
function of the time t, this function being graphically
represented by the current curve, which may be of any
shape. The effective value of the alternating current is
thus I_0, and is given by the equation

$$I_0 = \sqrt{\frac{1}{T} \int_0^T I^2 \, dt}.$$

The question we have now to consider is whether the
ampèremeter will show this or some other value. All
the instruments used for measuring alternating currents
are based upon some electro-dynamic or some heat effect
produced by the current. In either case the momentary
action is proportional to the square of the momentary
value of the current; and all these instruments will therefore be equivalent as regards their suitability for measuring alternating currents. We may thus restrict the
investigation to one particular type of instrument; the
result is valid for all other types.

For this purpose let us select the well-known dynamometer invented by Weber. The movable coil is suspended within the field of the fixed coil, and is thus
subjected to a deflecting force, which is proportional to
the square of the current, and which is balanced by the
torsional force of a spiral spring. In measuring a continuous current, I_0, the deflecting force is given by the
equation $I_0^2 \frac{1}{K^2}$, K being a constant depending on the con-

struction of the instrument. The balancing force exerted by the spring is proportional to the angular deflection D, which must be given to the torsion head to keep the pointer at zero. We have thus the equations

$$I_0^2 = K^2 D$$
$$I_0 = K \sqrt{D},$$

which is the well-known formula showing the relation between current and reading on a dynamometer.

The question we have to consider is whether the same formula is applicable if the current is not continuous, but alternating according to any irregular or regular function of the time. Let this function be represented by a curve, in which time is measured on the horizontal, and current strength on the vertical. By squaring the ordinates we obtain a second curve, and the area enclosed between the axis of abscissæ and this second curve represents the expression $\int_0^T I^2 \, dt$, whilst the height of a rectangle of equal base and area represents the square of the effective current. The movable coil of the dynamometer is acted on by two forces; one being the tension of the spring, and the other the electro-dynamic force of the current. The former is constant, and the latter variable, but always opposed to the former. If I_t is the current at time t, the force acting at that time upon the movable coil is $D K^2 - I_t^2$. This force produces an acceleration given by the expression $\dfrac{D K^2 - I_t^2}{m}$, if by m we denote the mass of the movable coil reduced to the radius at which the force acts. If this mass were sufficiently small, then, this acceleration, which is alternately positive and negative, would produce a visible oscillatory movement of the coil, the velocity of movement being represented by the formula

$$v = \int_0^t \frac{(DK^2 - I_t^2)}{m} \, dt$$

whereby we assume that at time $t = 0$ the velocity is also zero. The mass of the movable coil is, however, very large in comparison with the forces acting on it, and the frequency with which the acceleration changes from a positive to a negative maximum is so great that no visible oscillation is produced in the coil. This, moreover, is an obvious condition for an exact measurement. The velocity v must at all times be infinitely small, that is to say

$$\int_0^T \frac{(DK^2 - I_t^2)}{m} \, dt = 0$$

From this condition we obtain

$$DK^2 \int_0^T dt = \int_0^T I_t^2 \, dt$$

$$DK^2 = \frac{1}{T} \int_0^T I_t^2 \, dt.$$

The expression on the right-hand side of this equation is obviously nothing else than the height of the rectangle previously mentioned, that is to say, the square of the effective value of the current. We have therefore

$$DK^2 = I_0^2,$$

and this is a proof that the dynamometer does really measure the true effective value of the current, whatever may be the shape of the current curve. The constant K of the instrument may once for all be determined by calibration with a continuous current, and then used for all measurements with alternating currents, provided always that the instrument does not contain any metal masses in such proximity to the movable coil that a disturbing effect through eddy currents is produced.

and movable ... is at ...

The wattmeter. — We have now to investigate the problem how a dynamometer may be used for measuring power, and to determine whether such measurements are accurate for currents of irregular shape. The arrangement of the instrument is shown in Fig. 69. *C* is the

Fig. 69.

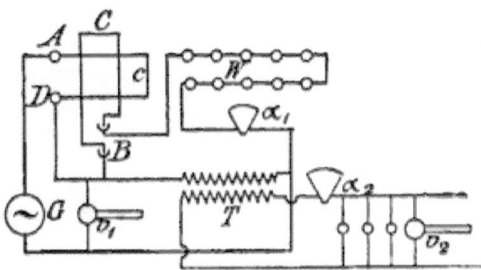

movable, and *c* the fixed coil. *G* is the alternator supplying current to the apparatus *T*, which may be a transformer.

The dynamometer, as usually constructed, has two terminals, *A* and *B* (or three if the fixed coil is arranged in two parts, so that the range may be increased). If the instrument is intended for measuring power, then the connection between the movable and the fixed coil must be provided with a third terminal, *D*. The current at which the power is to be measured is then led through the fixed coil, whilst a shunt containing an inductionless resistance, *W*, is arranged between the terminal *B* of the movable coil and the return circuit. The shunt resistance may consist of a ribbon of platinoid, constantan, or other alloy having a small temperature co-efficient, and the ribbon should be wound zigzag fashion, so as to reduce the reactance to a negligible quantity. A series of glow-lamps may also be used instead of a metallic resistance, if the relation between current and resistance has been previously ascertained. In the shunt circuit we insert an ampère-meter, a_1, although this is not absolutely necessary.

If the resistance W is very large, we may without sensible error assume that the shunt current is in phase with the E.M.F. impressed upon the circuit, which is shown by the voltmeter v_1. In other words, the current flowing through the movable coil has no lag, and is proportional to the E.M.F. The main current in the fixed coil may lag or lead as compared to the E.M.F. If the apparatus receiving power is a transformer, the current will lag, and the lag will be the greater the more reactance the apparatus has. Let I be the main current (maximum value I_m), and i is the shunt current (maximum value i_m), then if E.M.F. and current follow a sine law, the turning moment acting on the movable coil is proportional to

$$I_m \sin (a - \phi)\, i_m \sin a$$

a representing the phase of the E.M.F. at the time to which the expression refers, and ϕ the angle of lag of the current passing through c. If by W we denote the resistance of the shunt, including that of the ampèremeter a_1 and connections, we have

$$i_m = \frac{c_m}{W},$$

and the turning moment may also be considered proportional to

$$I_m \sin (a - \phi) \frac{c_m}{W} \sin a.$$

By twisting the torsion-head of the instrument we apply a twisting couple to the movable coil, which is given by the expression $D K^2$, D being the angular deflection, and K the constant of the instrument. Let m be the mass of the movable coil reduced to the radius at which the forces act, then the acceleration at the time to which the phase angle refers is

$$\frac{D\,K^2 - I_m\,\frac{e_m}{W}\sin(a-\phi)\sin a}{m}$$

and the velocity attained after the time t is

$$v = \frac{1}{m}\int_0^t \left(D\,K^2 - I_m\,\frac{e_m}{W}\sin(a-\phi)\sin a \right) dt,$$

provided at time $t = o$ the velocity is zero. The mass of the coil is very great as compared with the forces, and the changes in direction of the electro-dynamic force occurs with great rapidity, so that the movable coil cannot follow these changes, and remains at rest. In other words, v is at all times zero, and this requires that

$$t\,D\,K^2 = \int_0^t I_m\,\frac{e_m}{W}\sin(a-\phi)\sin a$$

This equation may also be written as follows—

$$D\,K^2 = \frac{I_m e_m}{W}\int_0^{2\pi}\sin(a-\phi)\sin a\,\frac{d\,a}{2\,\pi}$$

The value of the integral is $\frac{\cos\phi}{2}$ and we have therefore

$$D\,K^2 = \frac{1}{W}\,\frac{I_m e_m}{2}\cos\phi$$

or by introducing the effective values of current and E.M.F.

$$D\,K^2 = \frac{1}{W}\,I\,e\cos\phi.$$

It has already been shown that the product $I\,e\cos\phi$ is the power carried by the current I flowing under a potential difference e, and having a lag ϕ. The product

$W D K^2$ is therefore the power of the alternating current. The symbols have the following meaning—

D is the angular deflection of the torsion-head,

K is the constant of the instrument which has been determined by calibration with a known continuous current i, so that $K = \dfrac{i}{\sqrt{D}}$,

W is the total shunt resistance in Ohm.

We may use the instrument as above explained with reference to Fig. 69, and also as an ordinary dynamometer, using the terminals A and B, but not D. We measure first the main current I, arranging the connections as shown in Fig. 70, and then the shunt current $e \mathbin{/} W$ arranging the connections as shown in Fig. 71. If it is permissible

Fig. 70.   Fig. 71.

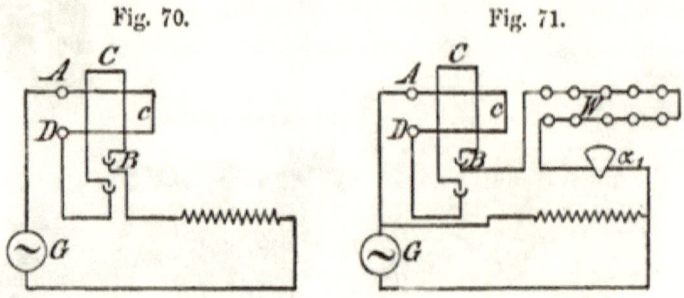

to neglect the self-induction of the movable coil as compared with the self-induction of the apparatus *T*, and if it is also permissible to neglect the resistance of the fixed coil as compared with the resistance of the shunt circuit W, then the dynamometer does not increase the lag nor waste power. Let the reading, when arranged according to Fig. 70, be D_1, and when arranged according to Fig. 71, be D_2, then

$$K\sqrt{D_1} = I,$$

$$K \sqrt{D_2} = \frac{c}{W}.$$

By multiplying these equations we obtain

$$K^2 \sqrt{D_1 D_2} = I \frac{c}{W}.$$

Previously we obtained the expression

$$K^2 D = I \frac{c}{W} \cos \phi$$

which, combined with the above, gives

$$\cos \phi = \frac{D}{\sqrt{D_1 D_2}}.$$

By thus taking three measurements with the same instrument (one power measurement and two current measurements), we may find the lag ϕ. It is important to note that only the readings enter into the equation for the lag. In order to determine the latter, it is therefore not necessary to know the constant of the instrument, nor need we know the shunt resistance.

Measurement of the power carried by currents of irregular form.—Up to now we have assumed that current and E.M.F. follow a sine law. If they follow any other law, the measurement taken by the wattmeter, as explained with reference to Fig. 69, will still be perfectly correct. The work done by the current in time T is

$$\int_0^T I c \, dt = W \int_0^T I i \, dt.$$

The main current I, as well as the shunt current i, may be any periodic function of the time, the only condition being that the frequency must be the same for both. This is obviously the case, since both are produced by the same E.M.F. The turning moment acting on the movable coil is $D K^2 - I i$, and varies continuously. If the frequency were sufficiently small (T very large), and the mass of the coil also very small, the coil would assume an oscillatory movement. But the frequency is great, and the mass of the coil is so considerable that no movement takes place; in other words, the integral of the oscillation taken over the time of a complete period, is zero.

$$\int_0^T \left(\frac{D K^2 - I i}{m}\right) dt = 0$$

$$T D K^2 = \int_0^T I i \, dt$$

$$W D K^2 = \frac{W}{T} \int_0^T I i \, dt.$$

It was shown above that $\dfrac{W}{T}\int_0^T I i \, dt$ is the work done by the current in time T. Since power is work divided by time, we have

$$W D K^\nu = P,$$

that is to say, the reading of the wattmeter gives the true power P, whatever may be the shape of the current and E.M.F. curves.

To make a power measurement by means of a wattmeter, we must know the constant of the instrument, and the exact value of the shunt resistance W. If the latter consists of platinoid or other alloy of small temperature co-efficient, this condition causes no difficulty. If, however, a series of glow-lamps is taken as the shunt resistance, the value of the latter is not constant, but varies with the E.M.F. To find W we may proceed in either of two ways: Read the deflection D on the wattmeter, the current i on the ampèremeter a_1, and the E.M.F. e on the voltmeter v_1. The two latter readings give $W = e / i$, and the power is then given by the expression

$$P = \frac{c}{i} D K^2.$$

The difficulty of this method lies in the necessity to take three readings simultaneously. To obviate this, we may do away with the reading of the current by calibrating the resistance beforehand, which may be done with continuous currents. A curve may be plotted, giving W as a function of c, and then it suffices to take the readings of D and c, W being taken from the curve. The ampère-meter a_1 may be cut out of circuit, whereby the conditions of an inductionless shunt is more easily fulfilled.

Fig. 72.

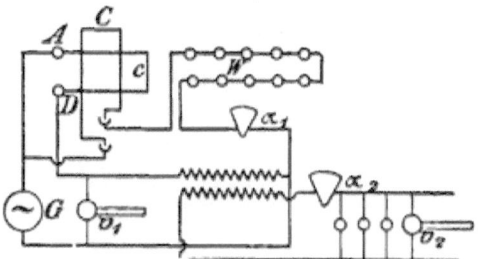

If the instrument is connected as shown in Fig. 69, it shows the power supplied to the transformer and that wasted in the shunt, that is, the total power supplied by the generator. If only the power given to the transformer is to be measured, the wattmeter must be connected up as shown in Fig. 72, where the shunt is taken from the wire which leads to terminal A. In this case, the fixed coil c carries only the current passing through the primary of the transformer, and the power wasted by the current in the movable coil C is not measured. The expression

$$P = W D K^2$$

gives then exactly the power supplied to the transformer.
If the load on the latter consists entirely of glow-lamps,
then the product of secondary current and secondary
terminal pressure is the power delivered by the transformer. We measure on the ampèremeter a_2 the secondary
current i_2, and on the voltmeter v_2 the secondary pressure
c_2. The efficiency of the transformer is then given by the
expression

$$\eta = \frac{i_2 c_2}{W D K^2}.$$

If the supply voltage is high, it is advisable to so con-

Fig. 73.

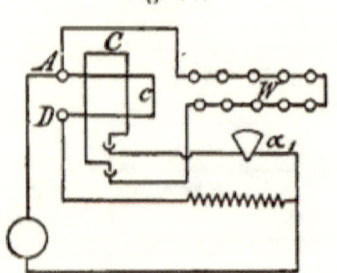

nect the wattmeter that in the instrument itself no great
potential difference can arise. Otherwise there is the
risk of breaking down its insulation. Fig. 73 shows the
arrangement of connections which should on that account
be avoided. Theoretically Fig. 73 is equivalent with Fig.
72; the latter arrangement is, however, from a practical
point of view, preferable, because the highest potential
difference which can arise between the fixed and movable
coil is only that due to the resistance and reactance of the
latter, and is therefore only a small fraction of the total
pressure. On account of safety in handling the instrument,
it is also advisable to earth that terminal of the generator

which is directly connected with the terminal *A* of the wattmeter.

The theory of the wattmeter given above is only correct if the assumption of a perfectly .inductionless shunt circuit is justified, by reason of the inductionless resistance being very great as compared with the self-induction of the movable coil in series with it. There will then be almost no lag of the current in this coil. To reduce the lag absolutely to zero is, of course, impossible, since the action of the instrument pre-supposes the existence of a

Fig. 74.

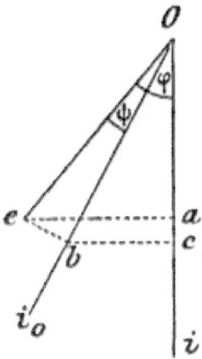

mechanical force, which can only be obtained by means of a coil producing a magnetic field of its own, that is to say, having a certain amount of self-induction which must produce some lag. With careful workmanship, the self-induction can, however, be brought down to such limits that the condition $i = \dfrac{e}{W}$ is practically fulfilled. If this condition is not fulfilled, a correction to the reading must be made as shown in the following theory.

Let *Oe* in Fig. 74 represent the vector of the total

supply pressure, and Oi that of the main current through the fixed coil, which has a lag ϕ, so that

$$\tan \phi = \frac{2\pi \sim L}{W},$$

$2\pi \sim L$ is the reactance, and W the resistance of the apparatus to which power is supplied. The power is $Oi \times Oa$.

If there were absolutely no self-induction in the watt-meter, the shunt current would be in phase with Oe. As there is some self-induction, it lags behind by an angle which we denote by ψ, and assumes in the diagram the position Oi_0. If w is the shunt resistance, and l its coefficient of self-induction, we have

$$\tan \psi = \frac{2\pi \sim l}{w}.$$

The wattmeter will then not indicate the true power $Oi \times Oa$, but the apparent power $Oi \times Oc$. To find the true power, we must therefore multiply the reading by the ratio $\frac{Oa}{Oc}$. Let P' be the apparent power shown by the instrument, and P the true power, then

$$P = P' \times \frac{Oa}{Oc}.$$

Since $Oa = Oe \times \cos\phi$ and $Oc = Ob \times \cos(\phi-\psi) = Oe \times \cos\psi\cos(\phi-\psi)$ we have also

$$P = P' \frac{\cos\phi}{\cos\psi\cos(\phi-\psi)}.$$

Since ψ is a constant of the instrument, it may be determined once for all. The phase difference between main and shunt current may be found from

$$\cos(\phi-\psi) = \frac{D}{\sqrt{D_1 D_2}}$$

as previously explained. Since ψ is known, we know also ϕ, and can determine the correcting factor. The above expression may also be written in the form

$$P = P' \frac{1 + \tan^2 \psi}{1 + \tan \phi \tan \psi}.$$

Is $\psi < \phi$, the correcting factor becomes smaller than unity, that is to say, the true power is smaller than that read off on the wattmeter; is $\psi > \phi$ (which may happen if the power circuit has capacity, or only resistance, whilst the wattmeter has too much self-induction), then the true power is larger than that read off on the wattmeter. There are two values of ψ, for which the reading of the wattmeter gives absolutely the true power. This will obviously be the case if the correcting factor equals unity; namely if $\psi = 0$, and $\psi = \phi$. In the former case we have an instrument with negligible self-induction, and in the latter case the self-induction in the shunt happens to be of such a value as to produce the same lag as in the main circuit. The case most frequently occurring in practice is that there is some lag in the main circuit, and that $\phi > \psi$ especially with wattmeters of modern construction, where ψ is very small. The correcting factor is then slightly less than unity, and attains its minimum if ϕ happens to be equal to 2ψ. This gives the greatest error possible, and the correcting factor is then $\dfrac{\cos \phi}{\cos^2 \dfrac{\phi}{2}}$. For all other values of ϕ, either greater or smaller than 2ψ, the error is less, and the correcting factor approaches more nearly to unity. The following table gives the maximum possible error for different values of ϕ.

ϕ	Power reading on Wattmeter	True power	The error is equal to or smaller than
5°	1000	998·5	·15 per cent.
10°	1000	994·6	·54 ,, ,,
15°	1000	982·7	1·73 ,, ,,
20°	1000	968·8	3·12 ,, ,,
25°	1000	950·8	4·92 ,, ,,
30°	1000	928·2	7·18 ,, ,,

Other methods of measuring power.—Measurement of power may be made in a variety of ways. For convenience and accuracy the wattmeter is preferable, but where such an instrument is not to hand, some other method must be employed. We may broadly divide all power measurements into two classes : (*a*) Absorption methods, whereby the power to be measured is absorbed by the apparatus itself which we use for measuring ; and (*b*) Transmission methods, where the power merely passes through the measuring apparatus, but is not absorbed therein.

As an example of the first kind of measurement, we may take the arrangement often used for alternators, whereby the output of the machine is taken up in a resistance which need not be absoluely inductionless. It should, however, consist of a metal or alloy having a small temperature co-efficient, such as platinoid, manganin, constantan, etc. For the measurement we require an accurate ampèremeter and a Wheatstone bridge. After running long enough to get the resistance heated up to its final temperature, a reading of the ampèremeter is taken, and the current is then switched off the resistance, the time being carefully noted. As quickly as possible the resistance is connected to the Wheatstone bridge, and resistance measurements are taken at frequent time-

intervals. A curve may then be plotted giving the resistance as a function of the time, and this curve, pro-
longed backwards to the moment when the machine current
was switched off, gives the actual value of the resistance
when the current was passing through it. The power
taken up is then given by the product resistance × square
of current. By adjusting the bridge beforehand, and
arranging the switches so as to facilitate the changing
over from the power circuit to the bridge circuit, a high
degree of accuracy may be attained by this method.

Another example of the absorption method is the
measurement of the power wasted in a transformer by
means of temperature readings. The temperature of the
transformer is taken at normal load; the alternating
current is then switched off and a continuous current is
sent through the primary, which is so regulated as to keep
the temperature up to the same point. The power wasted
by the continuous current is the same as that previously
wasted in the transformer, working under normal conditions, and the latter can therefore be accurately determined by volt and current measurements. It is advisable
to use a spirit thermometer, since a mercury thermometer
may show too high a reading, the error being due to eddy
currents set up in the mercury in the bulb by the stray
magnetic field. On the whole this method of measuring
power cannot be recommended; it is at best only a crude
approximation, and requires much time and personal skill.

As examples of the transmission method of power
measurement, may be mentioned the use of the wattmeter
as already explained, the "Three-Voltmeter Method" of
Professor Ayton, and the "Three-Ampèremeter Method"
of Professor Fleming.

Three-voltmeter method.—The application of this

method will be understood from Fig. 75, where G is the generator, a an ampèremeter, W an inductionless resistance,

Fig. 75.

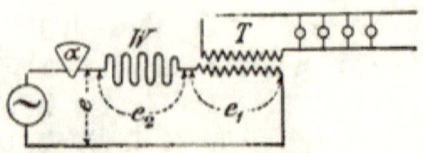

and T the transformer absorbing the power which we wish to measure. A voltmeter is placed between the main leads; let its reading be c. Another voltmeter is used to show the potential difference e_2 between the terminals of the inductionless resistance, and a third instrument shows the potential difference between the terminals of the primary of the transformer. Instead of using three separate voltmeters, we may of course use the same instrument for taking all three readings by arranging a suitable system of switches. This arrangement is preferable on account of its greater simplicity, and because slight errors in the calibrations of the instrument have less influence on the result. The clock diagram of this arrangement is shown in Fig 76. OI is the current, $OE_1 = e_1$ is the E.M.F. impressed on the transformer, and E_1E is the E.M.F. absorbed in the resistance W. Since the latter is inductionless, its vector E_1E must be parallel to the current vector OI. $OE = c$ is the total E.M.F. The watt-component of the impressed E.M.F. e_1 is $e_w = OA$ and the energy is $OI \times OA$. Since only the coil has inductance we have $AE_1 = BE = e_s$, the E.M.F. of self-induction, and the following equations obtain—

$$e_s^2 = e_1^2 - e_w^2.$$
$$e_s^2 = c^2 - (e_w + e_2)^2,$$

from which we find

$$c_w = \frac{c^2 - c_1^2 - c_2^2}{2\, c_2}$$

The power is given by the formula—

$$P = i\frac{c^2 - c_1^2 - c_2^2}{2\, c_2}.$$

To find the power we must take four readings, namely, three voltmeter readings and one ampèremeter reading.

Fig. 76.

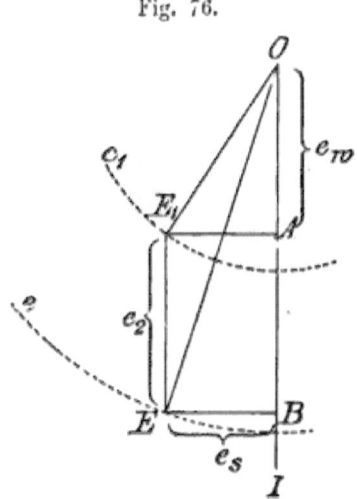

If the resistance W is accurately known, the last reading may be omitted and the power calculated according to the formula—

$$P = \frac{c^2 - c_1^2 - c_2^2}{2\, W}.$$

This is the power actually supplied to the apparatus T, in our case a transformer. If we want to know the power supplied by the generator G, e_2^2/W, the power lost in the resistance must of course be added, and we obtain

$$P' = \frac{e^2 + e_2{}^2 - c_1{}^2}{2\,W}.$$

Instead of calculating P, we can find the watt-component of e_1 graphically by drawing circles with radii e_1 and e, and shifting a vertical line parallel to itself until a position is found in which the piece contained between the two circles is exactly equal to e_2. This gives the position of the point E_1 in Fig. 76, and therefore also the length of the vector $OA = e_w$. The power is then

$$P = i\,e_w.$$

The diagram shows at a glance that a small error in the volt-measurements will produce the larger an error in the determination of the power the nearer the circle of e_1 is to O or e, and that the error will be least if e_1 is midway between O and e. To obtain an accurate measurement of power by this method we must, therefore, so chose the resistance W that e_2 does not sensibly differ from e_1, that is to say, that about the same pressure is lost in the resistance as is used in the apparatus under test. The total voltage e must then be considerably greater (from $1\frac{1}{2}$ to 2 times) than that required by the apparatus under test. If a suitable generator is at hand, or if we can transform up to get this extra voltage, the method is convenient and accurate, but failing these conditions the method is inapplicable. In this case we can use the analagous

Three-ampèremeter method, devised by Dr. Fleming.— This method is especially applicable if the testing current is obtained from an electricity works at the pressure required by the apparatus to be tested. The arrangement is shown in Fig. 77. The current is supplied at the terminals K and passes through an ampèremeter on the other side of which it is divided into two circuits, one containing the transformer T to be tested, and the other an inductionless resistance W. These two currents are measured on the ampèremeters a_1 and a_2; the pressure is measured on the voltmeter c. The clock diagram of this combination is shown in Fig. 78, where OE represents

Fig. 77.

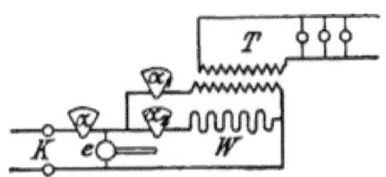

the pressure of the supply current i, the primary current of T and i_w its power component; i_2 is the current flowing through the resistance W; and its vector must of course be parallel with OE. From the diagram it will be seen that the following relation obtains—

$$i_1^2 - i_w^2 = i^2 - (i_w + i_2)^2$$

The power is given by

$$P = c i_w = \frac{c}{2} \frac{(i^2 - i_1^2 - i_2^2)}{i_2}$$

If the resistance W is accurately known, the reading for c

Fig. 78.

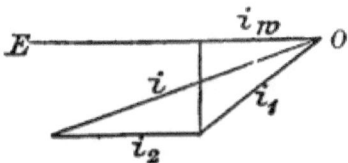

need not be taken, and the power may be calculated from

$$P = \frac{W}{2} (i^2 - i_1^2 - i_2^2).$$

Also in this method accuracy depends upon the proper choice of the resistance. It should be so adjusted that i_2 is not sensibly different from i_1; the total current i will then be from $1\frac{1}{2}$ to 2 times the primary current i_1 taken by the transformer.

In considering both methods, we have tacitly assumed that current and pressure follow a sine law; the question now arises, whether these methods will give accurate results if this condition is not fulfilled, that is to say, if the curves representing E.M.F. and current are of irregular shape. That the wattmeter gives correct indications also in such cases has already been shown, and since simultaneous measurements by means of a wattmeter and one or the other methods here described are always in accord, we naturally conclude that these methods must also be generally applicable. Apart from such experimental proof, this can also be shown by theory. For this purpose we shall consider the three-voltmeter method, the application to the analogous case of the three-ampère-meter method will then be self-evident. Let in the following the letters e and i denote the instantaneous values of E.M.F. and current respectively, then the expression

$$e = e_1 + e_2$$

is valid at any time for t. We also have at all times

$$i = \frac{e_2}{W}$$

and the power at any moment is

$$p = i\,e_1 = \frac{e_1 e_2}{W}\,;\ e_1 e_2 = p\,W$$

$$e^2 = e_1{}^2 + 2e_1 e_2 + e_2{}^2$$

$$e^2 = e_1{}^2 + 2p\,W + e_2{}^2$$

$$p = \frac{e^2 - e_1{}^2 - e_2{}^2}{2\,W}$$

The work done in the time T of a complete cycle is $\int_0^T pdt$, and the effective power is

$$P = \frac{1}{T} \int_0^T pdt$$

$$P = \frac{1}{T} \frac{1}{2W} \left(\int_0^T e^2 dt \ - \int_0^T e_1^2 dt - \int_0^T e_2^2 dt \right)$$

It has been previously shown that the expression $\frac{1}{T} \int_0^T e^2 dt$ is simply the square of the effective pressure indicated by the voltmeter; if now we denote these effective pressures by e, e_1, e_2 respectively, we have

$$P = \frac{1}{2W}(e^2 - e_1^2 - e_2^2)$$

Since in arriving at this result (which is exactly the same as that reached by the graphic method), we have made no assumption whatever as regards the shape of the E.M.F. curve, it follows that the three-voltmeter method is applicable to currents of any form.

Testing transformers.—By means of the various methods above explained the output and efficiency of transformers can be determined. It is of course necessary to have a source of current capable of supplying all the power wanted, and some apparatus capable to absorb the full output of the transformer. To obtain by this direct method anything like a reliable figure for the efficiency, input and output must be measured with extreme accuracy, the reason being that the two are not very different, and a small error in the determination of one or the other causes a great error in their calculated ratio. Let for instance the real input be 100 and the real output 97 Kwt., and let there be an error of 1 per cent. in each measurement, the error being negative in the measurement of the input and positive in the measurement of the output. The measurements would then be 99 and 98 Kwt. respectively, and the calculated efficiency would be 99 per cent. instead of 97 per cent., which it really is. To reduce as much as possible the magnitude of the error in the determination of the efficiency, it is advisable to make this determination by an indirect method in the following way. The test is made simultaneously on two equal transformers, which are so connected that the output of No. 1 forms the input of No. 2, and the output of this, supplemented by an external source of power again, the input of No. 1.

Fig. 79.

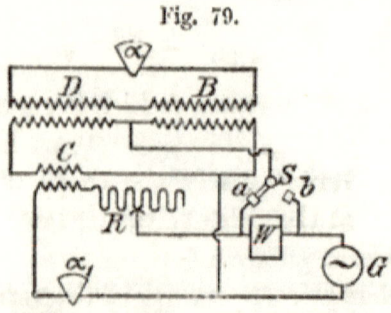

We obtain thus a circulation of power through the two transformers, and need only supply as much power as is wasted in both. This is a small amount, and need only be measured with a moderate degree of accuracy. The power circulating is also measured, and it will be obvious that small or moderate errors in both measurements cannot seriously affect the accuracy of the result. The arrangement of apparatus is shown in Fig. 79. *D* and *B* are the two equal transformers, and *C* is a small auxiliary transformer which supplies the waste power and thus keeps the total power in circulation. Into the primary of *C* we insert an inductionless rheostat *R*, for the purpose of adjusting the pressure supplied to C, so as to obtain in the ampèremeter a the normal full load secondary current of the big transformers. The connection between the latter must of course be so arranged that their E.M. Forces oppose each other. If the large transformers were only connected to C, the full current could be obtained in them, but not the pressure. To insure that also the right pressure is maintained in B and D, we connect their primaries with the generator G, as shown in the diagram. If now we open the rheostat completely, and short-circuit the secondary of C, then the generator has to supply only the no load losses of B and D, which will be indicated on the wattmeter W, provided the two-way switch S is put to contact a, as shown in the diagram. Since both transformers are equal, no current will be indicated in a. Now let us insert C and adjust the rheostat until a indicates the full load current. Then the large transformers are both working under full load and the wattmeter measures all losses. These are : The ohmic loss in R and the losses in C, B and D. Since we know the efficiency of the small transformer, and can measure its primary current on the ampèremeter a_1, we can calculate the power which the small transformer supplies to the two large transformers. If the switch is placed on its contact a, the wattmeter indicates the total power given off by the generator. Let this be P_1. If the switch is placed on b it indicates only that part of the power which flows to the rheostat and auxiliary transformer. Let this be Pc. The primary current of C is measured in a_1. Let this be i. If w is the resistance of the rheostat, corresponding to the position of its contact, then the power supplied to C is $P_c - i^2 w$. Let η' be the efficiency of C, then $\eta' (P_c - i^2 w)$ is the power given off by C. The generator delivers to the primaries of the two transformers the power $P_1 - P_e$ and the auxiliary transformer delivers $\eta'\ (P_e - i^2 w)$. The total power wasted in B and D is therefore

$$P_v = P_1 - P_e + \eta'\ (P_e - i^2 w).$$

If P is the output and η the efficiency of transformer D, then D receives an input of P watt and gives an output of $\eta\ P$ watt. B receives an input of $\eta\ P$ watt, and gives an output of $\eta^2\ P$ watt. From this follows

$$P_v = P - \eta^2 P$$
$$\eta = \sqrt{\frac{P - P_v}{P}}.$$

Since P_v is very small as compared with P, it is obvious that a moderate error in the determination of P_v can only produce a very small error in the determination of the efficiency η. According to our previous example, we would have for $P = 100$, $P_v = 6$. If we measure accurately, we obtain

$$\eta = \tfrac{1}{10}\sqrt{94} = 0.9695.$$

Now let us assume that we do not measure accurately, but commit an error of 1 per cent. in the measurement of P and 5 per cent. in that of P_v, then the maximum possible error which can occur in the determination of the efficiency is only $\frac{1}{4}$ per cent.

　This indirect method of determining efficiency is therefore far and away more accurate than the direct method. It has moreover also other advantages. In the first place, no apparatus is required for taking up the output of the transformer under test; and, secondly, only a moderate amount of power need be employed. Both are important considerations when large transformers are to be tested.

The heating of transformers can be investigated in the following way. The transformers are first heated in a stove or drying-room to that temperature which they will probably assume in continuous work. Where a hot room is not available, the heating may also be done by passing through the fine wire winding a continuous current from a secondary battery. When the transformers have been brought up to near their probable working temperature, they are loaded, the connection of **Fig. 79** being preferably employed, and the temperature is taken at stated time-intervals by means of spirit thermometers. The readings are plotted in a curve of which the abscissæ represent time and the ordinates temperature, and the test is continued until the curve becomes horizontal. If an uninterrupted supply of alternating current is available, the preliminary heating can be dispensed with. The transformers are then connected up from the first, as shown in Fig. 79, and kept at work until they have attained their maximum temperature.

The drop may be tested directly under normal working conditions, or by the indirect method given in Chapter VI. The latter is more simple and accurate.

The insulation test should be taken immediately after the heating test. It is also advisable to flash the transformer, so that any weak spot in the insulation may be found out and remedied before the apparatus is set to work. For this purpose, temporary connections should be made between (*a*) a primary and secondary terminal; (*b*) a primary terminal and earth; (*c*) a secondary terminal and earth. Care must of course be taken that during these tests both poles of the generator are well insulated from earth.

Testing sheet-iron.—The methods in use for the deter-

mination of the permeability and the hysteresis loop of
sample bars cannot conveniently be applied for sample
sheets, because the attainment of a really reliable magnetic
contact at the ends of the sample is very difficult. It is
of course possible to find the hysteresis loop for sample
sheets prepared as rings, but since each ring has to be
specially wound with a magnetizing and a pilot coil, this
method is too cumbersome for practical work. Moreover,
the result is arrived at indirectly. From a practical point
of view we do not require to know the exact shape of the
hysteresis loop; all we care to know is the power wasted
both in hysteresis and in eddy currents in a known weight
of sheet-iron at a given induction and a given frequency.
The most direct method is obviously to measure the power
lost in the sample by means of a wattmeter. The permeability cannot be found that way, but can be estimated
with a fair degree of accuracy from the power factor of
actual transformers when working with open secondary.
Transformers used for this purpose must of course not have
butt-joints.

A very simple way for testing sheets is to make stampings of the shape used for transformers, and to insert them
in the usual way into a coil of known number of turns.
The induction is found from the terminal pressure, the
section of iron S inside the coil and the number of its
turns n.

$$E = 4\cdot44 \sim n S B 10^{-\kappa}.$$

The ohmic loss of pressure being very small may be
neglected. The total loss is due to hysteresis and eddy
currents in the iron. If the induction remains the same
whilst the frequency is changed, then the loss through
eddy currents varies with the square of the frequency,

whilst the hysteresis loss varies as the frequency. It is thus possible to separate the two losses. For this purpose we need only vary E proportionally with $\sim$ and determine the power lost in each case. Let us assume that we have made two measurements, P_1 and P_2 for the power lost, which correspond to the two frequencies $\sim_1$ and $\sim_2$ respectively. We have then

$$P_1 = h \sim_1 + f \sim_1{}^2$$
$$P_2 = h \sim_2 + f \sim_2{}^2$$

h and f being as yet unknown co-efficients relating to hysteresis and eddy current losses. These co-efficients can now be found from the two equations, and we obtain thus a means to separate the two losses. The hysteresis loss for the frequency $\sim$ is then given by

$$P_h = h \sim_1$$

and the eddy current loss is given by

$$P_f = f \sim_1{}^2$$

In testing iron as here described, it is of course necessary to employ the sample sheets in a form which permits the building up of a transformer core. It is however not always convenient to stamp or cut-out the sheets in this particular form, and if a method can be found by which the samples may have the shape of straight strips the preparation of the batch would be easier, and no iron need be wasted.

Such a method of testing has been devised by several engineers. Herr von Dolivo Dobrowolsky uses the apparatus shown in Fig. 80.[1] It consists of two ⌐⌐ shaped cores of sheet-iron, which can be laid together either directly or placed on either side of the sample $A\,A$ to be

[1] This illustration is taken from the *Elektertechnische Zeitschrift*, 1892, No. 30, page 406.

tested. The sample is composed of rectangular sheets and forms the common yoke to the electro-magnets *n s.* When the magnets are placed directly in contact, the direction of the current through the coils is such that both drive the induction in the same sense; when the sample is inserted the connections are charged by means of the switch *B,* in such manner as to produce the polarity indicated in the diagram. The flux now passes from both magnets through the yoke. The current is measured by

Fig. 80.

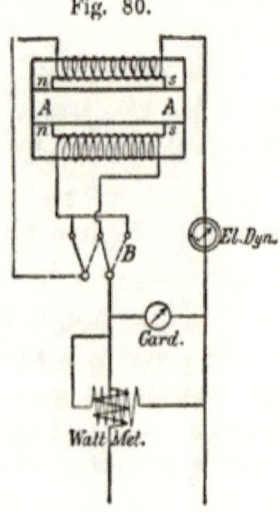

a dynamometer marked *El. Dyn.* in the figure, and the pressure by a Cardew voltmeter marked *Card.* The power is measured by a wattmeter inserted as shown. In using the apparatus the magnets are laid together and the switch is put into the position which produces circular magnetization. The power corresponding to various values of the induction is then measured, the induction being calculated from the frequency, the pressure and the known data of the coils and magnet cores. The sample is then inserted,

the switch changed over and the measurements repeated.
The sectional area of the sample should be about double
that of the magnets. The difference between the two
sets of measurements is then the power wasted in the
sample at the various values of the induction. A drawback of this method is the difference in megnetic leakage
with and without the sample. If the magnets are laid
together directly, and magnetized circularly, there is hardly
any leakage, and B can be calculated from E with great
accuracy. If the sample is inserted, the magnetic resistance
is increased, and leakage produced which diminishes the

Fig. 81.

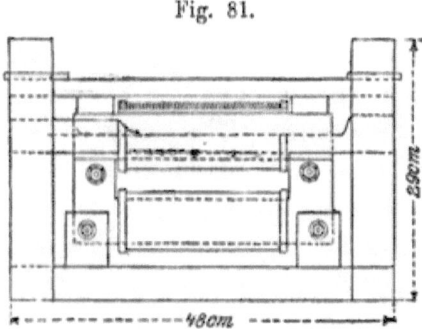

value of B in the sample. At the same time there is a
difference in the value of the induction along the magnet
cores, the induction being a maximum in the centre of
each core. E can therefore no longer be regarded as an
exact measure for B, and an error is thus introduced.

To avoid this difficulty, the author has constructed the
apparatus shown in Fig. 81. The sample consists in this
apparatus also of a batch of rectangular plates, and forms
one of the two longer sides of a rectangular frame, the
three other sides being formed by ⌊⌋ shaped plates of
known magnetic quality. Both larger sides are sur-

rounded by coils, the upper one being large enough to admit the insertion of the sample without difficulty. The connection is made for circular magnetization, so that only very little leakage takes place, and this is the same for all samples. The sample must have approximately the same cross-section as the magnet. To calibrate the instrument, a sample is prepared from the same iron as the magnet, and after weighing the total amount of iron in the magnet, the loss of power is determined for different values of B. This loss is then allotted between magnet and sample according to their relative weights, and a curve is plotted showing the loss in the magnet as a function of B. If now another sample is inserted, and the total loss measured, we have only to deduct from it the loss as found from the curve for the particular induction observed, and the rest is the loss in the sample. The curves of Fig. 9 have been partly plotted from tests made with this apparatus, and partly from tests made on actual transformers.

Another apparatus for factory use has been devised by Prof. Ewing. Its principle is the purely mechanical determination of the loss in a sample of very small dimensions, namely 6 to 8 strips of 3 in. length and $\frac{5}{8}$ in. width. The apparatus consists of a permanent magnet e, Fig. 82, which is suspended on knife edges f, and weighted by a screw g. For transport the magnet can be raised off the knife edges by means of a rack and wheel h. A dashpot below the magnet serves to steady its swing, and a pointer moving over a scale at the top shows the deflection produced when the sample a is rotated between the poles. The sample is fastened by screw clamps $b\,b$ to a carrier which can be rotated by means of a handle, and the friction wheels $d\,c$. The screw i serves to level the instrument. The reversal of magnetism in the sample is produced by the rotation, and the work lost in hysteresis and eddy currents per revolution is $2\pi \times$ Torque. The torque is indicated on the scale by the pointer, and since 2π is a constant, we find that the deflection of the pointer gives directly a measure for the loss per cycle, the speed of rotation

Fig. 82.

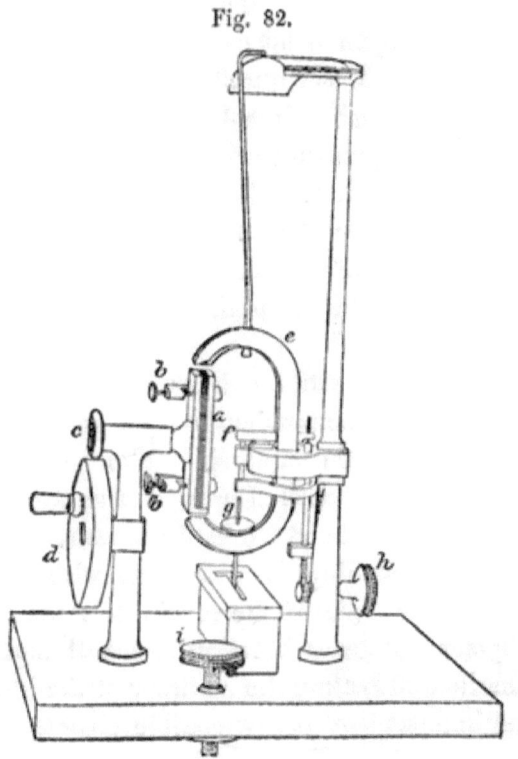

having no influence as long as it is not so high as to sensibly augment eddy current losses.

The sample sheets are prepared to a gauge, the length being sensibly less than the polar gap of the magnet, so that the magnetic resistance of the air gap preponderates

over that of the sample itself. The object of this arrangement is to avoid the error which might otherwise be
introduced when samples of widely different permeability
are tested. The magnet produces in the sample an induction of about 4000 CGS units, but this can be slightly
raised or lowered by taking less or more sample plates.
Prof. Ewing found that an accurate adjustment as regards
the weight of samples is not required, since the deflection
varies but slightly if the number of plates making up a
sample batch is varied. It suffices to adjust the weight
of the batch roughly to that which corresponds to 7 strips
of 13 to 14 mils. thickness. This is the thickness frequently
chosen for transformer sheets. When testing armature
plates, which are usually stouter, a correspondingly smaller
number of strips would be used to make up the sample
batch.

The apparatus is calibrated by using samples, the
hysteresis of which has previously been accurately determined by the ballistic method. Two such standard
samples are supplied with the apparatus, together with
tables giving the results of the ballistic tests. In testing
other samples, a reading is also taken with one of the
standards, and the ratio of the readings is taken as the
ratio of hysteresis losses between standard and sample.
By this method of testing, the accuracy of the instrument
is rendered independent of any possible change that may
have occurred in the strength of the permanent magnet.

CHAPTER VIII

Safety appliances for transformers.—The reason why we use transformers is that we may carry the power under high pressure, and distribute it under low or moderate pressure. It is, however, an essential condition that the insulation between the transmission circuit (primary) and the distributing circuit (secondary) be absolutely perfect. If this condition be not fulfilled the advantage of transforming becomes illusory, and the use of transformers may even become dangerous on account of an unjustified feeling of security. The two windings in a transformer must necessarily lie in close proximity, and thus an injury to the insulation may cause a leakage of current and a transfer of pressure from the primary to the secondary coil. Since in a widely distributed net-work of primary conductors their insulation cannot be absolutely perfect, it will be obvious that any leak between the primary and secondary coil of any particular transformer may raise the absolute potential of the secondary to a dangerous amount. This potential will depend on the position of the leak in the transformer, on the position of the equivalent leak in the general system of high pressure or primary circuits and on the insulation of the secondary circuit. It may be a few hundred volts only, or it may be equal to the full primary voltage. If in the latter case a person touches any part of the secondary circuit he will receive a dangerous or fatal shock. To avoid this danger several expedients are possible. One very obvious preventive is to place between the two windings a metallic dividing-sheet which is well earthed. If the insulation between the two windings is damaged, contact is not made between the primary and secondary direct, but through the intervention of this dividing-sheet, and thus the potential of the secondary is prevented from rising to any dangerous extent. This appliance ensures safety only in so far as regards a leak from one winding to the other, but it is useless against a leak in any other part of the transformer, for instance, between the primary and secondary leading-in wires. Although in a proper design, and with careful workmanship, such a leak may be regarded as almost impossible, yet we must admit the possibility that design or fitting up may not be perfect, and for such cases protectionary measures become necessary. The simplest way of obtaining protection is to earth some point of the secondary circuit, preferably the middle of the winding, since then the potential difference of the secondary mains to earth becomes a minimum, namely, equal to half the voltage. If contact takes place anywhere between primary and secondary, the former is thereby connected to earth, and all danger of a fatal shock is avoided. The danger as regards fire is, on the other hand, increased by this expedient. If the whole of the secondary circuit is insulated from earth, a fault must occur at two places of different potential before a danger in respect of fire can arise, but if one point of the secondary circuit is permanently con-

nected to earth, a fault occurring in one place only is sufficient to create danger. The margin of safety is therefore
reduced by one half if we earth the middle of the
secondary winding.

This drawback is overcome by the safety appliance
introduced a few years ago by the Thomson-Houston
Company. The appliance consists of an earth-plate and
two metal knobs *a*, *b*, Fig. 83, which are connected to
the secondary mains. Between the knobs and the earth-
plate is inserted a thin sheet of insulating material (paraffined paper or mica). As long as no fault between

Fig. 83.

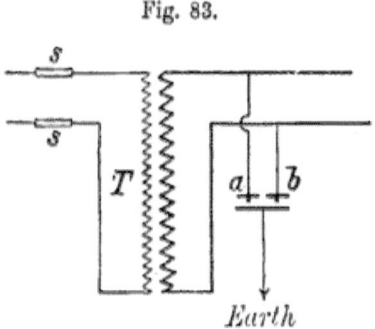

Earth

primary and secondary occurs, the potential difference between the knobs and the earth-plate remains within the
limit of the secondary voltage, and this is not sufficient to
break down the insulation between knobs and earth. If,
however, through a fault in the insulation between
secondary and primary, the secondary assumes the potential of the primary, the insulation between *a* and earth
and *b* and earth is broken down, thereby short-circuiting
the secondary winding. The primary current then rises to
such an amount that the safety forces *s s* go, and the
transformer is thereby automatically cut out of circuit.

N

A safety device invented by Major Cardew, and much used in England, is shown in Fig. 84. In this arrangement the action depends on electrostatic attraction between a plate E connected to the secondary, and an aluminium foil lying on a plate connected to earth. The aluminium foil has the form of two discs connected by a narrow bridge, and is together with the two plates enclosed in a box, provision being made by means of a screw thread in the cover of the box to accurately adjust the distance between the plate E and the aluminium foil. The latter is

Fig. 84.

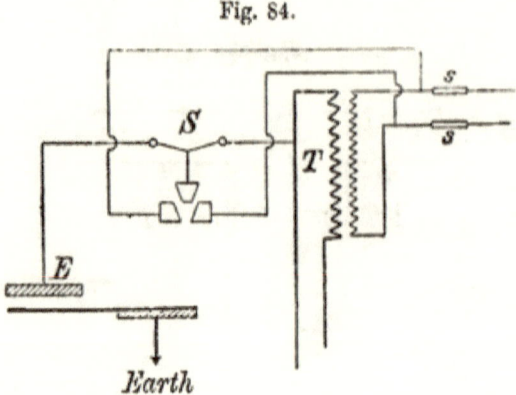

Earth

permanently kept at the potential of the earth (zero), whilst the plate E has under ordinary circumstances a potential not exceeding the secondary voltage. The electrostatic attraction corresponding to this potential difference is insufficient to raise the foil; if, however, a fault occurs between primary and secondary the potential difference immediately rises to such an amount that the electrostatic attraction suffices to raise the foil and bring it into contact with the plate E, thereby earthing the secondary winding. In the safety device first described

by Cardew[1] a fuse S was provided and arranged to hold up
a weight which, if the fuse melted, would short-circuit the
primary leads, and thus cause their fuses s s to go, and
the transformer to be cut-out of circuit. It has, however,
been found that this is a superfluous refinement, since the
short produced on the secondary by the lifting of the
aluminium foil is in itself sufficient to make the primary

Fig. 85.

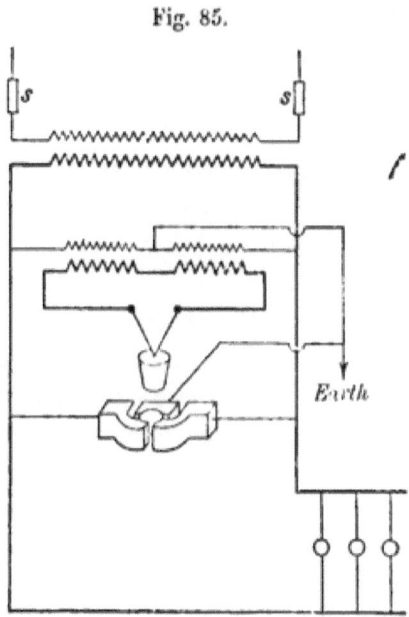

fuses go. The apparatus can be set to come into action
if the potential of the secondary rises to 400 volts. Hence
even an incipient fault in insulation between the two
circuits is sufficient to automatically disconnect the
transformer from the circuit.

Ferranti's safety device is shown in Fig. 85. The

---

[1] *Journal Inst. El. Eng.*, vol. xvii. p. 179.

secondary mains are connected to the primaries of two very small transformers coupled in series, whilst their secondaries are coupled in parallel. The secondaries are connected to a fuse carrying a conical weight over a corresponding set of terminals. The connection between the two primaries is joined to earth, as is also one of the terminals, the other two being joined to the secondary mains. As long as the insulation between the primary and secondary circuits of the main transformer is perfect, there is absolute balance between the E.M. Forces of the secondary windings of the two small auxiliary transformers, and no current passes through the fuse. If, however, a fault occurs, the balance is disturbed, a current passes through the fuse and melts it, and the weight falling between the terminals short-circuits the secondary mains, and puts them to earth. The primary fuses *s s* are thereby caused to blow, thus cutting the faulty transformer completely out of circuit. It is important to note that this safety device is a protection, not only against a real short between primary and secondary, but even against an incipient fault of insulation between the two circuits.

Sub-stations and house-transformers.—The most important use of transformers is in connection with lighting or power plants, where the pressure has to be raised or lowered. The nature of the glow-lamp and the condition of personal safety make it necessary to use a moderate pressure in the distributing leads (say not exceeding 100 or 200 volts), whilst a high pressure in the transmission leads is an economic advantage, and, indeed, an absolute necessity, if the transmission has to be effected over a considerable distance. The transformer is then the intermediary apparatus by which the two conditions, cheap mains and moderate supply voltage, can be simultaneously fulfilled. The typical arrangement of transformers for this purpose is shown in Fig. 86. *C* denotes omnibus bars in the central station; *S s* the primary transmission mains or feeders; *T T* are transformers, and *V V* the supply mains. Measuring instruments, switches, and fuses are of course also required, but have been omitted from the diagram to avoid complication.

Fig. 86.

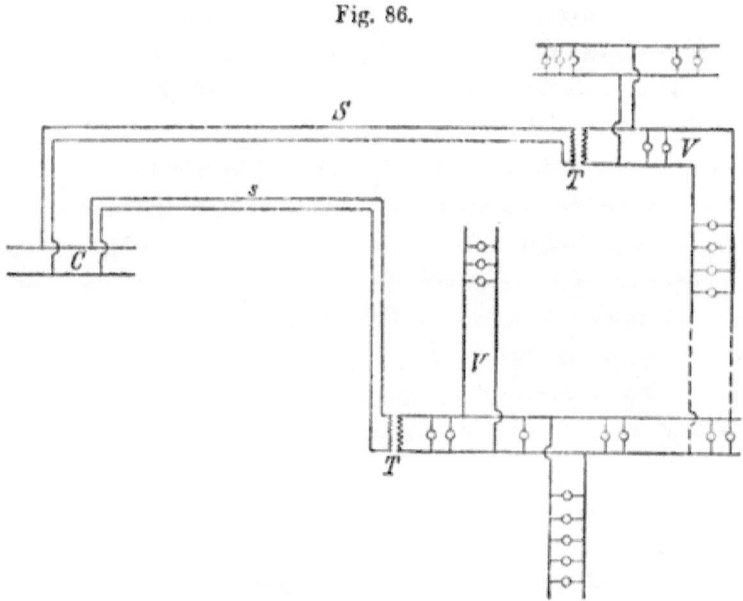

The diagram shows each transformer supplied with current by its own feeder, whilst on the secondary side each transformer supplies a net-work of distributing mains, which latter may be either separate from each other, or they may be inter-connected, as shown by the dotted line. The inter-connection of secondary mains has the advantage that at times of small demand some of the transformers may be disconnected from the primary and secondary mains, whereby the power wasted by them when working an open circuit is saved. On the other hand, there is the danger that a defect in one of the small net-works may affect all the others, and to minimize this danger it is advisable to insert fuses into all the important junctions of the secondary net-work. A system arranged as here described is called a *sub-station system*, and is characterized by the use of high-pressure feeders, and a complete net-work of low-pressure distributing mains.

The system of house-transformers, on the other hand, is characterized by a complete net-work of high-pressure feeding and distributing mains, supplying current to a large number of small transformers, each placed as near as possible to the place where the low-pressure current is required (*i. e.* one transformer to each house or group of houses), so that no net-work of secondary or low-pressure street mains is required. The weight of copper in the street mains is thereby much reduced, which is an advantage. On the other hand, there are some drawbacks. Owing to the greater length and the many junctions in the system of high-pressure mains, the insulation is more difficult, the high-pressure must be brought into the houses of the consumers, and the loss of power in the transformers is greater. Single transformers cannot be disconnected, thus increasing the light-load loss, and even at heavy load the loss of power is greater, since small transformers cannot have as high an efficiency as large transformers.

We may show the relative merits and faults of the two systems by an example. Take a district in which 100,000 50-watt lamps are installed. If we use house-transformers their joint output must be 5000 Kwt. Although all the lamps installed will never be alight simultaneously, it may

and will occasionally happen that all the lamps installed
in one house are simultaneously in use, and the output of
the transformer must therefore be equal to this demand.
This necessitates the employment of a large number of
small transformers (say from 2 to 10 Kwt. output) which
have collectively an output of 5000 Kwt. If sub-stations are
used we may take advantage of the fact, that of all the
lamps installed in a town only a certain percentage is in
use simultaneously. This percentage varies according to
the character of each district, but we may take 60 per cent.
as a very ample allowance. The sub-station transformers
need then only have an output of $5000 \times 0.6 = 3000$ Kwt.,
and may each be of a fairly large size, since only few of
them are required for this total output. As has already been
shown in Chapter V., the annual loss by ohmic resistance
is very small as compared with the hysteresis loss. We
may, therefore, without committing any great error, neglect the copper loss for the present, and compare the two
systems as regards iron loss only. As a fair average we
may take $3\frac{1}{2}$ per cent. iron loss for the small house-transformers, and 2 per cent. for the larger transformers at the
sub-stations. We shall also assume that in neither case
transformers are cut-out of circuit during the times of
light load.

Assuming 600 hours as the average annual burning time
of each lamp, we would then have an annual output of
$3000 \times 600 = 1.8 \times 10^6$ Units. If house-transformers are
used, the loss going on day and night would be 5000
$\times 0.035 = 175$ Kwt., making in the course of a whole
year $175 \times 8760 = 1.53 \times 10^6$ Units. The total work
(neglecting ohmic loss) which must be supplied during
the year is, therefore,

$$(.18 + 1.53) \, 10^6 = 3.33 \times 10^6 \text{ Units,}$$

and the annual efficiency is

$$\frac{1\cdot8}{3\cdot33} = 0\cdot54$$

The efficiency is in reality slightly less than 54 per cent., because the loss due to ohmic resistance, which we neglected, must be also covered by the work supplied.

If sub-stations are used the loss is considerably less, namely, $3000 \times 0\cdot02 = 60$ Kwt., or in one year 525,600 Units. The work put into the transformers is thus

$$(1\cdot8 + 0\cdot526)\ 10^6 = 2\cdot326 \times 10^6 \text{ Units}$$

annually, and the annual efficiency is

$$\frac{1\cdot8}{2\cdot326} = 0\cdot779.$$

The efficiency is also in this case slightly lower because of the ohmic loss. If we allow for the latter 2 per cent. in both cases, we have the following comparison between the two systems—

	House-Transformers.	Sub-Stations.
Annual efficiency per cent.	52	72
Annual output in units ...	1,800,000	1,800,000
„ input „ „ ...	3,460,000	2,400,000
„ loss „ „ ...	1,660,000	600,000

If a unit costs 1.2*d.* to produce in the central station, the money loss chiefly occasioned by hysteresis in the transformers is £8300 and £3000 respectively. The difference, namely £5300, shows by how much the working expenses are increased if we employ house-transformers instead of sub-stations. On the other hand, there is a saving in capital outlay with the former system. To put the comparison on a proper basis, we must now investigate

by how much the capital outlay must be lessened in order
to make the system of house-transformers commercially
better than that of sub-stations. This will be the case if
the annual charges for interest, sinking fund, and mainten-
ance for cables and transformers show a difference of more
than £5300 in favour of the house-transformer system.
We may roughly estimate these charges at 10 per cent. of
the capital outlay for the transformers and 15 per cent. for
the cables. The capital outlay for small transformers may
be taken at £3 10s., and for large transformers at £2 15s.
per Kwt. The total cost of transformers is thus for the
system—

Of house-transformers £17,500, or £1750 annually.
Of sub-stations £8250, or £825 annually.

The difference of £925 being in favour of the sub-station
system has to be added to the £5300 saved in power, so
that the saving in the annual charges for the cables must
exceed £6225 before house-transformers are commercially
preferable to sub-stations. The difference in the capital
outlay for the cables in the two systems must (with 15 per
cent. annual charge) come to or exceed

$$\frac{6225}{0\cdot15} = £41,500$$

or a little more than 8s. per lamp installed. If then
estimates of the two systems show that with house-transformers we save in cables alone more than 8s. per lamp as
compared with sub-stations, then the former system should
be adopted. This consideration is, of course, only valid for
the particular percentage charges on which the calculation
has been based. If, instead of allowing 15 per cent. on
cables, we had allowed a smaller charge, we would have

obtained more than 8*s*. per lamp as the critical difference in the capital outlay for cables. The same would have been the case if we had estimated the works' cost of the unit at more than 1.2*d*. It will also be obvious that the cost of cables in both systems, and therefore the difference in the cost of cables per lamp installed, must be greater for smaller stations, especially if the lamps are distributed over a wide area. Bearing in mind these various points, we come to the following general principles.

The system of house-transformers is commercially advantageous under the following conditions—

> Power cheap.
> Lamps scattered over wide area.
> Cables dear in first cost and upkeep.

The system of sub-stations is commercially advantageous under the following conditions—

> Power dear.
> Station large.
> Lamps densely installed.
> Cables cheap in first cost and upkeep.

Boosters.—If some of the feeders between the central station and the sub-stations are very long, it is sometimes advantageous to allow a greater voltage drop in them than in the shorter feeders, and to raise the pressure at the home end of these long feeders by an amount corresponding to the extra drop. For this purpose special auxiliary transformers, so-called "boosters," may be used. This system of boosting-up the pressure at the home end of long feeders has been invented simultaneously and independently by Mr. Stillwell in America, and by the author in England. It is shown diagrammatically in Fig. 87.

C are the bus bars in the station, *S* is a feeder supplying current to the transformer *T* at a sub-station. *V*
are the distributing mains connected to this transformer.
The boosting transformer *B* has its primary permanently
connected to the bus bars, whilst its secondary is put in
series with the feeder and is subdivided into sections, so
that by using a switch *s*, a greater or lesser number of
secondary turns can be inserted. In this manner the
additional voltage put into the feeder at the home end
may be varied from zero to the full voltage given by all
the secondary turns of the booster. The full voltage is
added when the feeder carries its maximum load; the
switch is then placed on its highest contact. As the load

Fig. 87.

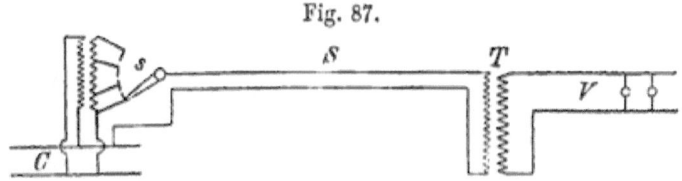

decreases the switch is shifted to a lower contact, the
intention being to boost up by the amount corresponding
to the drop in pressure due to ohmic resistance. Since
this drop is proportional to the current, the adjustment of
the switch may be made in accordance with the readings
of an ampèremeter in the feeder circuit, or pilot wires may
be brought back from the sub-station and connected to a
voltmeter. The switch is then adjusted so as to keep the
pressure indicated by the pilot voltmeter, constant. It
is obvious that in either case the switch-lever can be
worked automatically by a small electro-motor controlled
by a relais. Since, in passing from one contact to the
other, the switch-lever, if it were made in one solid piece,

would short-circuit, and possibly burn out the section of
the secondary winding connected to the two corresponding
contacts, it is necessary to employ a lever consisting of two
parts, each smaller than the width of the gap between two
contacts, and having an insulating partition between them.
The two parts must of course be joined by a suitable resistance, or preferably by a choking coil. With such a construction there can occur neither a short-circuit in the
booster nor an interruption of the feeder current.

The necessity to send the whole feeder current through
the switch, and the drawback of a complete interruption

Fig. 88.

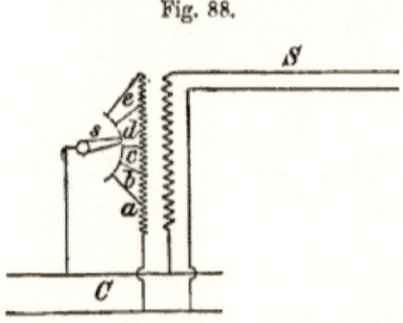

of the feeder current if this switch gets out of order, has
led the author to design the modified arrangement of
booster in which the switch is connected, not with the
secondary but with the primary circuit of the auxiliary
transformer. This arrangement is shown in Fig. 88. The
feeder circuit is permanently connected with the bus bars
through the secondary winding of the auxiliary transformer, whilst the multiple contact switch is inserted into
its primary connection with the bus bars. The primary
winding is subdivided into groups *a, b, c,* etc. According to
the position of the switch-lever, more or less of these groups
are active, thus causing the magnetic flux and the E.M.F.

in the secondary to be smaller or greater respectively. The first group a must of course contain a sufficient number of convolutions to prevent the auxiliary transformer from being magnetically overloaded. This kind of booster must therefore be larger than that shown in Fig. 87, but as in any case the cost of a booster is very small as compared with the saving in the cost of the feeder thereby rendered possible, the extra outlay is insignificant, whilst the possibility of keeping up the supply, even if the switch should become deranged, is a distinct advantage.

Fig. 89.

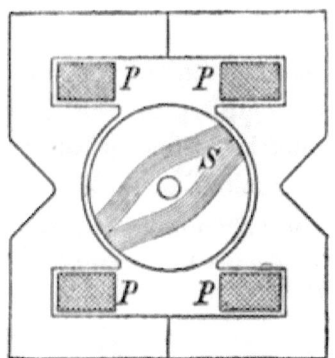

In a third type of boosting apparatus there is no switch of any kind, either in the secondary or primary circuit. This type is shown in Fig. 89. The construction resembles that of a two-pole dynamo with shuttle-wound armature. The field is built up of sheet-iron plates, and is provided with the primary winding P P, whilst the armature carries the secondary winding S placed over a core of sheet-iron discs in the usual manner. Both windings are permanently connected, the primary with the bus bars, and the secondary with bus bars and feeder as in Fig. 88. By means of worm gearing, the coil S may be placed at various angles with reference to the polar surfaces. If the coil S is turned into a vertical position, the flux of force passing through it is a maximum, and the E.M.F. generated in this coil is a maximum. If the coil be placed horizontally it is ineffective, whilst in intermediate positions any desired boosting effect may be obtained. By turning the coil beyond its horizontal position the action may also be reversed, that is to say, we can reduce the E.M.F. at the home end of the feeder. The advantages of this type of booster are that no switches of any kind are used and that the adjustment of the boosting effect is made, not by definite steps, but as gradually as we please, by means of the worm gear.

Connection in series.—Transformers may be advantageously used if it be required to work a number of lamps in series off a circuit in which an alternating current of constant strength is maintained. If we were to insert the lamps themselves into such a circuit, the insulation of the lamps to earth would have to be so perfect as to withstand the full potential difference of the alternating current, a condition not always easily fulfilled. If, however, we feed the lamps from the secondaries of series-transformers, it is only necessary to provide perfect insulation for the transformers, which presents no difficulty; the insulation of the lamps need only be good enough for the voltage required by each lamp. The arrangement is shown in Fig. 90. $T\,T$ are series-transformers supplied from a constant current alternator, and $L\,L$ are the lamps. The primary return circuit is not shown. Since the current in the primary is constant, the current in the secondary is also approximately constant as long as the lamp is in circuit. There is, however, the drawback that if a carbon should fall out of a lamp, or some other accident happen whereby the secondary current is interrupted, the induction in the core and the E.M.F. in the secondary of that particular transformer (if this is of the ordinary construction for parallel work) would rise very considerably. Since the primary current must, on account of the other lamps, be kept constant, the pressure at the generator has, in such a case, to be increased. The transformer with open secondary becomes magnetically overloaded and must eventually burn out. To avoid this danger we must make provision to give the secondary current an alternative path in case the lamp circuit should become interrupted. This may be done in two ways. We may employ a kind of automatic "cut-in"

Fig. 90.

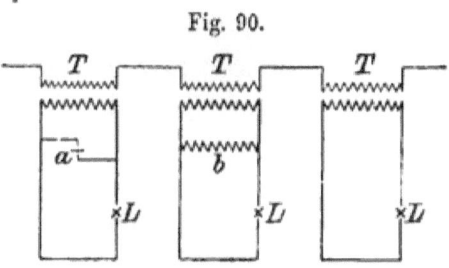

as in *a*, or a choking coil as in *b*. The cut-in consists of two electrodes separated by a thin sheet of mica or paraffined paper, which, under normal conditions, is sufficient to withstand the secondary voltage. If now the secondary voltage rises considerably, in consequence of the lamp circuit being opened, the insulation between the electrodes breaks down, and the cut-in short-circuits the secondary coil of the transformer. The choking coil, which may be used instead of a cut-in, is a kind of small transformer with only one winding on its core. The current passing through this winding is proportional to the lamp voltage and lags by nearly 90° behind it. The power lost in the choking coil is the sum of hysteresis and ohmic loss; it is of course not equal to the product of current and voltage, but much smaller. By a proper design of choking coil it is thus possible to minimize the loss of power, although current is apparently lost. This may best be seen by an example. Let us assume that the lamp requires 20 ampères at 35 volts, and that its power factor is 80 per cent. The power actually supplied to the lamp is therefore 560 watt. Let the choking coil be so constructed that it also takes 20 ampère if the pressure is 35 volts, and that the loss of power in it is 20 watt. Its power factor is therefore $20 : (35 \times 20) = 0.0285$. If we now draw a vector diagram to represent these working conditions, we find that the total secondary current supplied by the transformer is 36 ampère. If now the lamp current is interrupted the choking coil must pass the whole 36 ampère, and the voltage must rise to—

$$35 \times \frac{36}{20} = 63 \text{ volts.}$$

This is an excess of 80 per cent. over the normal voltage, and is accompanied by a similar rise in the magnetisation of the iron core. It is of course always possible to so design the choking coil that it can stand this increase of magnetic load without danger for any length of time.

Sometimes it is convenient to use a transformer for feeding a circuit of lamps in series, which requires a nearly constant current, although the number of lamps inserted may be varied. This condition is of course fulfilled if the primary current is constant, but if the primary voltage is constant a transformer for parallel work (that is, a transformer of the usual construction having as little magnetic leakage as possible) would be quite unsuitable. Such a transformer keeps the secondary voltage approximately constant, but not the secondary current. When we have lamps in series it is the current which must be kept constant, whilst the voltage must vary as nearly as possible in accordance with the number of lamps alight at any time. This condition can be met at least approximately by shaping the transformer in such way as to produce a large magnetic leakage. A construction of this kind is shown in Fig. 91. It is a core transformer with primary and secondary coils on separate limbs and with expansions a, b of the two yokes arranged specially to produce magnetic leakage. The primary coil is joined

Fig. 91.

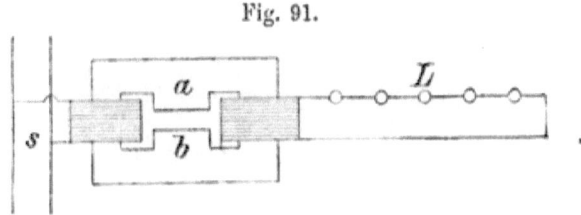

to the primary constant pressure lead s; and the secondary coil to the lamp circuit L.

It will be obvious that with an open secondary or lamp circuit the leakage field between a and b will be very small, since the core of the secondary coil offers a ready path for the magnetic flux. If, however, the lamp circuit be closed, a current flows in the secondary coil, pushing back part of the flux produced by the primary coil, and the leakage field, not only between a and b but all over the transformer, will be much increased. The larger the secondary current the more lines are pushed back, and the lower will be the secondary E.M.F. If a lamp is short-circuited the current will at first increase. This increase produces more magnetic leakage, and lessens thus the flux

o

which produces E.M.F. in the secondary. The increase in current strength will therefore be considerably smaller than would obtain with an ordinary transformer, and in this way it is possible to keep the current at least approximately constant when lamps are put out of action by being short-circuited.

The limits within which this kind of regulation is practically applicable can be seen from the vector diagram Fig. 92. Let OA be the current and OE_2 the secondary voltage at full load (all lamps alight). In the diagram the load is supposed to be without reactance. If E_2E_1 is the

Fig. 92.

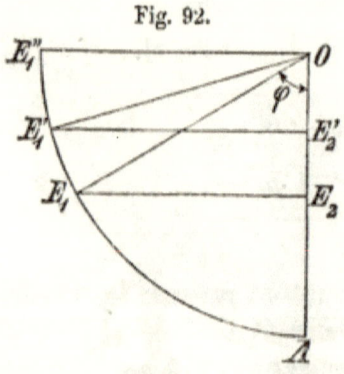

vector due to self-induction and resistance (the latter being assumed negligibly small) then OE_1 is the vector of the primary voltage. The length of the line E_2E_1 is of course proportional to the current. If now we short-circuit half the number of lamps, then only half the secondary voltage is required, and E_2 in the diagram moves to E_2', and E_1 moves to E_1'. The secondary current is now represented by the length of the line $E_2'E_1'$, that is to say, it is slightly larger than before. If we short-circuit all the lamps and the mains themselves, no E.M.F. is required, and E_2 moves to O, whilst E_1 moves to E_1''. The current is now given

by the line OE_1''. The ratio between the length of the
line E_2E_1 and OE_1'' shows the proportion of increase in
current strength if the load is reduced from the full
number of lamps to zero, and it is obvious that we can
make this ratio as near unity as we please, if we only
provide sufficient magnetic leakage, making the angle ϕ in
the diagram sufficiently large. The transformer must,
however, be made correspondingly larger, and will be more
costly, whilst its efficiency will necessarily be somewhat
smaller than that of a transformer designed for a minimum
of magnetic leakage. It is also important to note that
with arc-lamps, which have considerable self-induction,
this kind of regulation for constant current is only possible

Fig. 93.   Fig. 94.

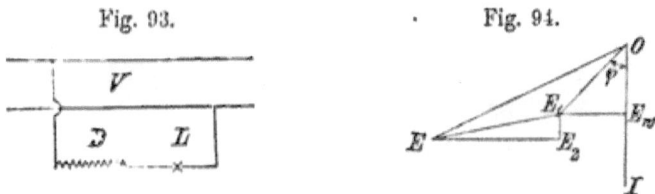

within narrow limits, since the line OE_2 is inclined to the
left, and the difference in the length of E_2E_1 and OE_1'' is
much greater than in Fig. 92.

Choking coils.—If lamps are worked in parallel from a
constant pressure circuit, the pressure at the lamp terminals may conveniently be reduced by choking coils. This
is preferable to the use of resistances, since very much less
power is thereby wasted. Let, in Fig. 93, V be a pair
of constant pressure mains and L a lamp which requires
a reduced pressure. This may be obtained by inserting a
choking coil D as shown. Fig. 94 is the corresponding
clock diagram. OI is the lamp current, OE_w the watt
component of the pressure, and OE_1 the total pressure at

the lamp terminals; the phase difference being given by the angle ϕ. For a glow-lamp ϕ would be zero and E_1 would coincide with E_{w}. With an arc-lamp there is a lag between current and E.M.F., making $OE_1 > OE_{w}$. The watt component of the E.M.F. in the choking coil is given by the vector $E_1 E_2$, and the wattless component by $E_2 E$. The vector $E_1 E$ is therefore the pressure which exists between the terminals of the choking coil and OE is the vector of the supply E.M.F.

Compensating coils.—The arrangement sketched in Fig. 93 may be used if an arc-lamp has to be connected to mains having a greater pressure than the lamp requires.

Fig. 95.

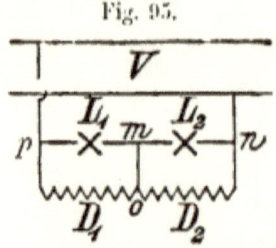

Alternating-current arc-lamps require from 30 to 35 volts at their terminals. It is thus possible to work three lamps in series from 110-volt mains. If only one lamp is required the excess of pressure must be taken up by a choking coil. Let us now assume that we have a pair of 65-volt mains, then two lamps may be worked in series. It is, however, not necessary to use both lamps together, for by inserting compensating coils, as shown in Fig. 95, we can render the lamps independent of each other. $D_1 D_2$ are the two equal windings of a transformer placed over the same core. In the diagram these windings are shown side by side for greater clearness. The coils are connected in o, and the direction of the winding is such

that a current flowing in D_1 from left to right produces
an E.M.F. in D_2 from right to left, and *vice versâ*. Imagine
now one of the lamps, say L_1, cut-out of circuit, then the
current will flow through D_1 to O, and finds there two
paths; the one through D_2 and n, and the other through
m L_2 n to the other main. It is obvious that the first
path is impassable to the current, since in D_2 acts an E.M.F.
which is always opposed to the direction of the current.
This E.M.F. produces in fact itself a current, which added
to the current coming from D_1 passes also through the
lamp. The two coils with their core act as a transformer
with the transforming ratio 1 : 1, D_1 being the primary
and D_2 the secondary winding. Let us assume, for the
sake of simplicity, that the efficiency of this transformer
is 100 per cent. If the lamps require each 12 ampère
then 6 ampère will pass through D_1 and 6 ampère will
pass in opposite direction through D_2, the two currents
joining at O and passing jointly through the lamp, which
thus receives a current of 12 ampère. At n the current
splits again, 6 ampère going to the other main V and
6 ampère going to D_2. Since the efficiency must be less
than 100 per cent., D_2 will contribute a little less and the
mains must contribute a little more than half the current
required by the lamp. The transformer has three
terminals, p, o, n, of which o is common to both coils and
is connected to the point m between the two lamps.

This principle of compensating coils may also be extended to more than two lamps. In Fig. 96 are shown
3 lamps and 3 coils. Let the lamp current be again
12 ampère and let the two lamps L_2, L_3 be switched out.
Then D_2 and D_3 will carry a current of a little over 4
ampère, which will produce in D_1 a current of a little under
8 ampère, so that L_1 is fed by the sum of these currents

and receives 12 ampère. These compensating transformers are often used in house installations, because they combine the advantage of the series arrangement of arc-lamps with the further advantage that the lamps are independent of each other. The output of these transformers is more-over less than that of a corresponding number of single

Fig. 96.

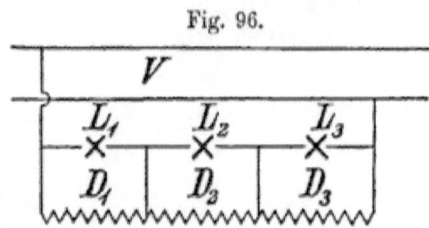

transformers, as will be seen from the following. Let e be the supply voltage, and P the power required by each lamp when the current is i. Then we require, for the arrangement shown in Fig. 95, an apparatus having an output of

$$e \times \frac{i}{2} = \frac{P}{2} \text{ Watt.}$$

We assume hereby, for the sake of simplicity, $\cos \phi = 1$. If we feed the two lamps by separate transformers, each would have to be designed for an output of

$$\frac{e}{2} \times i = \frac{P}{2} \text{ Watt.}$$

The combined transformer for two lamps in series contains no more material and costs no more than a single trans-former for one lamp.

If we use three lamps in series, the combined trans-former must be designed for the pressure e and the current $\frac{2}{3}i$. The amount of material required for its construction is, therefore, proportional to an output of $\frac{2}{3} P$ Watt, whereas the joint output of three single transformers would be $3 \times \frac{i}{3} \times e = P$ Watt. We see thus that also in this case there is some advantage in using a combined transformer.

Three-wire system.—If we connect the middle o of the secondary winding (Fig. 97) with a third main, we obtain a distributing system analogous to the so-called three-wire system used in connection with continuous-current stations. The primary coil pq receives high-pressure current from the feeder s, whilst the secondary coil mn is

Fig. 97.

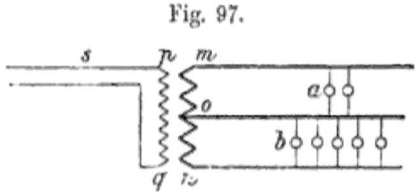

connected to the outer wires of the system in the usual way. The lamps a, b are connected between these outer wires and the zero wire o. The pressure between m and n is double the lamp voltage, and we are thus able, exactly as in the ordinary three-wire system, to effect considerable economies in the cost of the distributing mains.

Balancing transformer.—It may happen that the sub-station must be placed at some distance from the district to be lighted. In this case the middle wire need not be brought back to the sub-station transformer T, Fig. 98, if a balancing transformer T_1 is established in some point of the district to be lighted. The output of the balancing transformer need not be larger than half the maximum difference between the loads on the two sides a, b of the system. Let i_a be the maximum current in a and i_b the current which simultaneously obtains in b, then one coil of the balancing transformer must take up the current $\dfrac{i_a - i_b}{2}$ and its other coil must give off an equal current. (Compare also Fig. 95.) If the lamp voltage is e, then the output of the balancing transformer is given by the expression $\left(\dfrac{i_a - i_b}{2}\right)e$ the output of the sub-station transformer at the same time being $\left(\dfrac{i_a + i_b}{2}\right) 2e = (i_a + i_b)e$. Since it is,

Fig. 98.

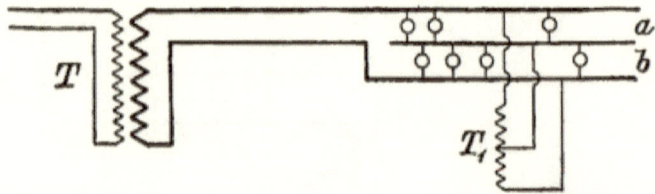

however, possible that both sides of the system may occasionally carry the maximum current, the sub-station transformer must be designed for an output of $2 i_a e$. If by p we denote the ratio of load difference between the two sides to the maximum load on one side, we have

$$i_b = (1 - p) i_a.$$

The output of the balancing transformer must therefore be

$$\frac{p\, i_a\, e}{2}$$

Since $i_a e$ is half the output of the sub-station transformer, we have the ratio between its size and that of the balancing transformer given by the fraction $4 : p$.

Thus for a load difference of 100, 50, 20, 10 per cent., the balancing transformer would be respectively $\frac{1}{4}$, $\frac{1}{8}$, $\frac{1}{20}$, $\frac{1}{40}$ the size of the sub-station transformer. These figures show that a comparatively very small balancing transformer may render it superfluous to carry the middle wire of the system back to the sub-station.

Scott's system.—An interesting application of transformers is the conversion of a two-phase into a three-phase system, and *vice versâ*, invented by Mr. C. F. Scott.[1] The arrangement is diagrammatically represented in Fig. 99, where G is a two-phase generator supplying current

Fig. 99.

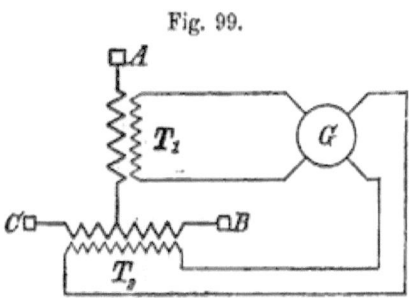

to the primaries of two separate transformers T_1 and T_2. The secondaries of these transformers are joined together and with the terminals A B C, as shown in the figure. Since the primary currents in T_1 and T_2 have a phase difference of 90°, there is also the same phase difference in the E.M. Forces generated in the two secondary coils. The E.M.F. between terminals A and B is therefore the resultant of two components, one being the full E.M.F. generated in the secondary of T_1, and the other half the E.M.F. generated

[1] *The Electrician*, April 6, 1894.

in the secondary of T_2, the latter component being moreover displaced by 90° as regards the former component.
Let, in Fig. 100, OA be the E.M.F. of T_1 and OB half
the E.M.F. of T_2, then BA is the resultant E.M.F. which
we measure between the terminals A and B. In the same
manner we find CA as the resultant E.M.F. produced by
T_1 and the left half of T_2, whilst CB is the E.M.F. produced by both halves of T_2. It will be obvious that, by

Fig. 100.

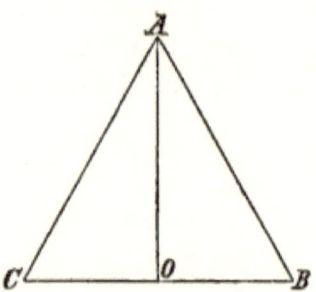

a proper choice of the number of turns in the secondaries,
we may so arrange matters that $OB = \dfrac{1}{2} AB$. Then AB
$= BC = CA$, and $OA = AB \sqrt{\dfrac{3}{4}}$, or $OA = 0\cdot867\ AB$
$= 0,0867\ BC$. The winding must therefore be such that
the secondary voltage of T_1 is $0\cdot867$ of the secondary
voltage of T_2. In the clock diagram the vectors of
terminal pressure pass then through zero at intervals of
60°, or in the same sense at intervals of 120°, which
characterizes a three-phase current. We obtain thus
from the terminals A, B, C a three-phase current.

The advantage claimed by Mr. Scott for this system
is that the generation and utilization of the current may

be effected by two-phase machinery, whilst the transmission may be made in three phases. The former
condition he considers to be an advantage as regards the
independent working of motors and lamps, and especially
their regulation, whilst the latter condition is, of course,
conducive to economy in copper on long lines of transmission.

A complete plant arranged according to Scott's system
is shown in Fig. 101. *G* is a two-phase generator pro-

Fig. 101.

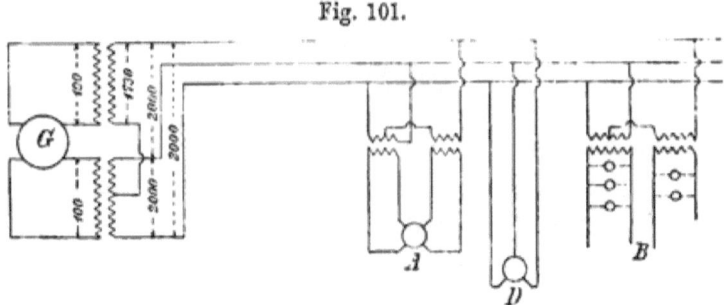

ducing 100 volt, which are transformed up to 2000 and
1730 volt in the two transformers shown. To the three
free terminals are joined the line wires, and between each
pair there is a pressure of 2000 volts. At the points of
consumption the three-phase current is either transformed
down and converted into a two-phase current for working
motors (*A*) or supplying light (*B*), or it may be used as
a three-phase current for working motors (*D*). Although
the circuits are inter-connected, the regulation for constant
pressure in the lamp circuits causes, according to the
inventor, no more difficulty than if the lamps were
connected directly with the generator.

CHAPTER IX

In order to give the reader a general survey over the best practice now obtaining in the design of transformers, a few examples of the apparatus built by leading firms are here given.

Messrs. Siemens and Halske of Berlin make core transformers for simple and multiphase currents. Fig. 102 shows a three-phase transformer with and without surrounding case. The cores are arranged as shown in Fig. 16, and are pressed against the yoke-discs by cast-iron heads with inclined shoes. To these heads are also secured the boards on which the terminals and fuses are attached. The leads are brought in through holes in the base-plate.

Fig. 103 shows a three-phase transformer with perforated iron case to give access to air for cooling. This type can of course only be used in dry places. If the transformer is to be erected in the open, a tight case, as shown in Fig. 102, must be used. For single-phase current the transformers are built similarly, but contain only two limbs. Figs. 104 and 105, in connection with the following tables, give the principal dimensions for the various sizes.

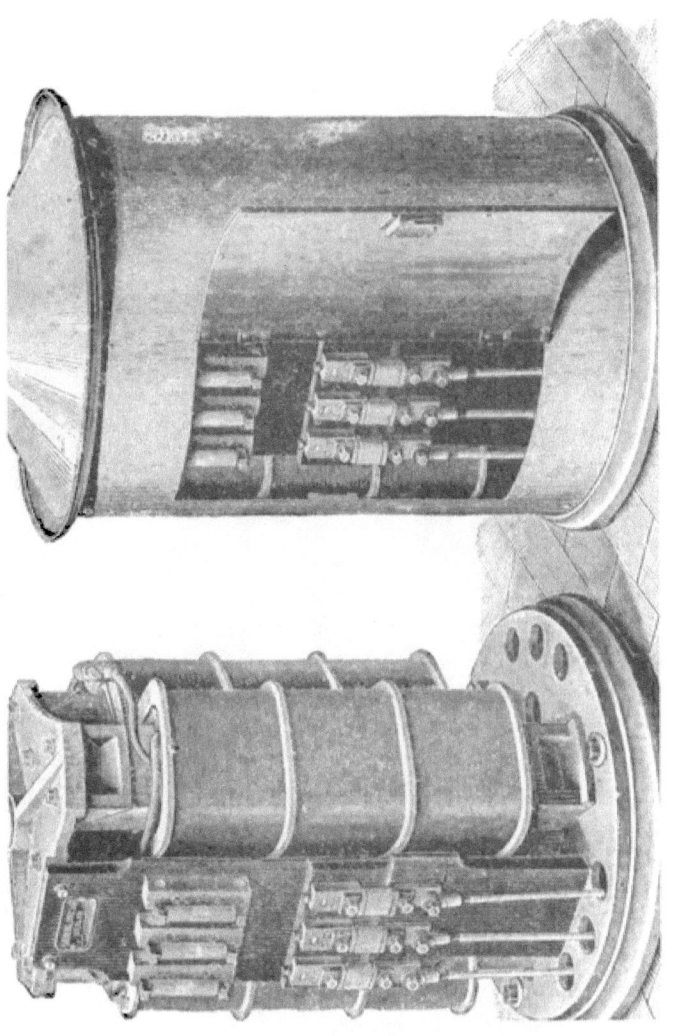

Fig. 103.

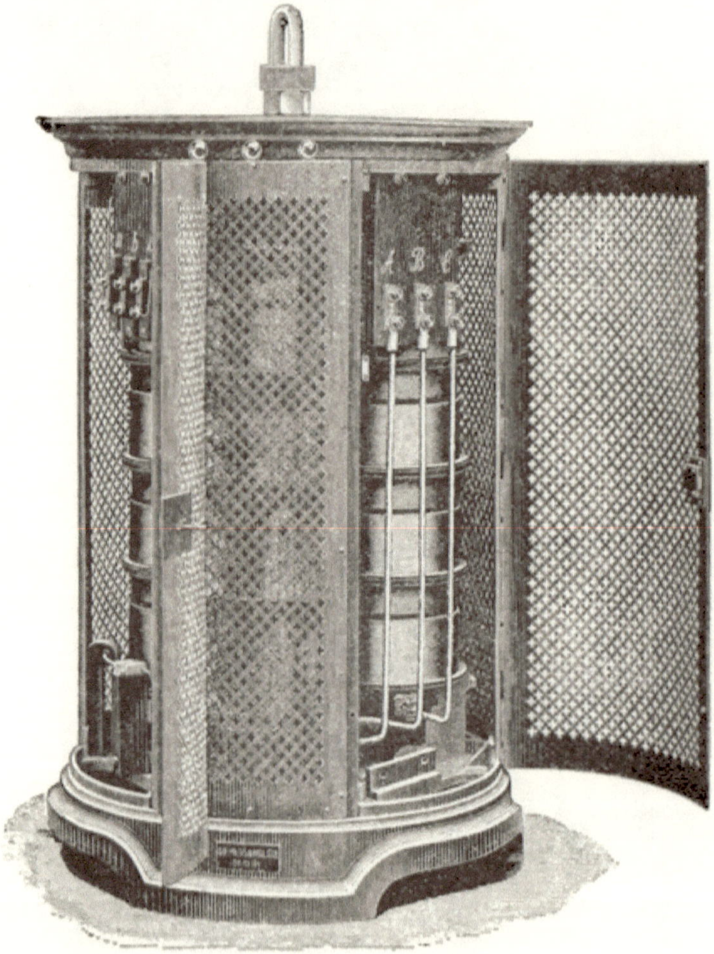

TABLE SHOWING DIMENSIONS OF THREE-PHASE TRANS-FORMERS IN MILLIMETRES.

Output. Kwt.	a	b	c	d	e	f	g
2·5	235	595	595	530	630	760	850
5	300	720	720	655	690	855	940
7·5	300	720	720	655	690	855	940
10	305	730	740	675	850	1020	1110
15	350	820	830	765	870	1065	1150
20	350	820	830	765	1070	1265	1350
30	405	980	1020	885	1250	1515	1660
50	425	1020	1060	925	1500	1730	1900
75	490	1160	1200	1065	1540	1855	2025
100	490	1160	1200	1065	1540	1855	2025
150	550	1300	1320	1185	1840	2155	2285
200	610	1400	1440	1305	2150	2455	2625

TABLE SHOWING DIMENSIONS OF SINGLE-PHASE TRANS-FORMERS IN MILLIMETRES.

Output. Kwt.	a	b	c	d	e	f	g
1	210	510	530	475	470	590	675
2·5	240	590	590	530	610	745	830
5	295	720	710	665	660	825	910
7·5	300	730	740	675	820	990	1075
10	350	820	830	765	870	1065	1160
15	350	820	830	765	1070	1265	1360
20	405	980	1020	885	1245	1500	1635
30	425	1020	1060	925	1500	1750	1880
50	480	1160	1200	1065	1540	1830	1960
75	480	1160	1200	1065	1540	1830	1960
100	540	1300	1320	1185	2070	2145	2280
150	610	1400	1440	1305	2140	2455	2285

Fig. 104.

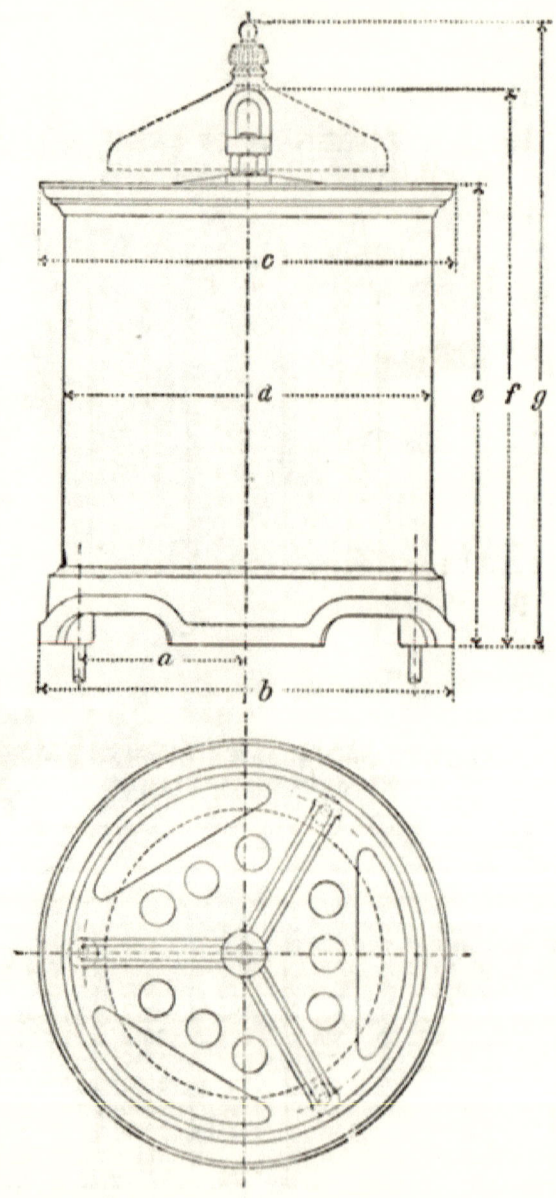

Fig. 105.

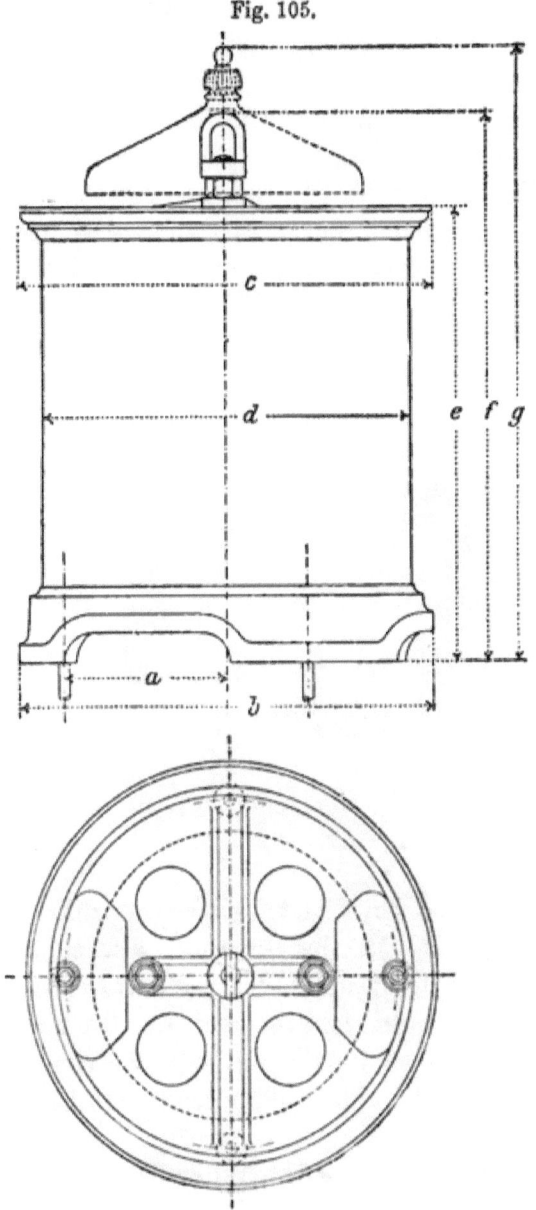

Messrs. the Elektrizitäts-Aktien-Gesellschaft vormals Schuckert & Co., of Nürnberg, build single-phase transformers of the shell type and three-phase transformers of the core type. As examples of the former we select a 2 Kwt. (Fig. 106) and a 40 Kwt. (Fig. 107), and of the latter a 10 Kwt. (Fig. 108). In the shell transformers the coils are bedded in iron, and at the ends they are protected by cast-iron covers. The latter are held together

Fig. 106.

by strong screw-bolts, whilst the shell is secured by angle-irons and bolts. In the three-phase transformers the coils are subdivided into a number of sections to minimize leakage. The apparatus is protected by a cylindrical outer casing, not shown in the figure. The connecting cables are brought through stuffing-boxes in the upper cover.

Messrs. the Maschinenbau-Aktien-Gesellschaft vormals L. Schwartzkopff, of Berlin, adopt for single-phase transformers the shell type with a long core (compare also Fig. 12^d), and increase the cross-section of the yoke, as compared to that of the core, so as to reduce hysteresis

Fig. 107.

losses. The iron parts are, as will be seen from Fig. 109, held together by strong cast-iron frames and screw-bolts. The coils are protected by perforated sheets. (Fig. 110.)

Fig. 108.

Fig. 109.

Fig. 110.

Messrs. Siemens Brothers and Co., Limited, of London, have also adopted the shell type, Fig. 12⁷, but avoid butt-joints by letting the full plate of one layer cover the joints in the next layer. The iron parts are held together by cast-iron frames and screw-bolts. (Fig. 111.) The coils are wound upon a cylinder of special insulating material

Fig. 111.

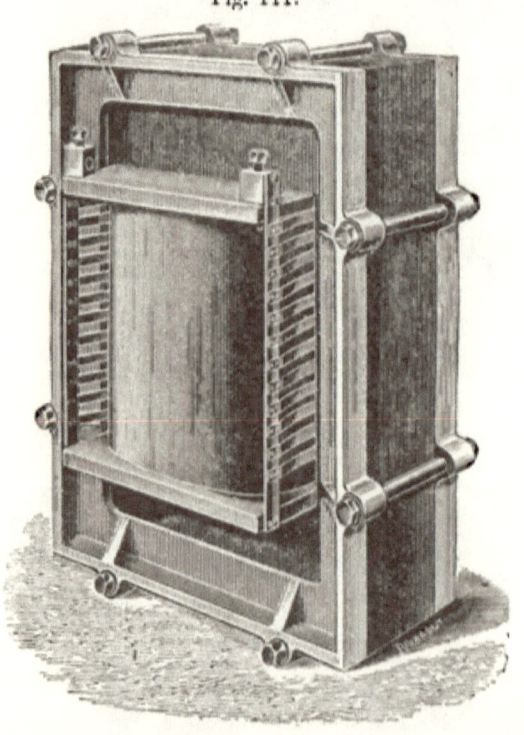

capable of withstanding heat, and which is provided with wooden flanges at both ends. The layers of the high-pressure winding are also separated from each other by thin sheets of special insulating material. The low-pressure winding consists in the transformer here illustrated of a number of wires wound in parallel and connected to

terminal bars which are attached to the wooden flanges.
The terminals of the high-pressure winding are also
attached to the upper flange, but on the other side and
by means of insulating washers of ebonite.

Messrs. Brown, Boveri and Co., of Baden (Switzerland),
have adopted a type of transformer which may be considered a compromise between the shell and core types.
The two coils are placed upon a single core, and there is
a yoke of |_____| shape to one side of this core.   Fig. 112
shows the details of this design, and Fig. 113 a complete
transformer.   The yoke-plates are driven tightly into the
main casting, the coils are slipped over the core, and the
latter is then laid across the yoke and pressed down into
it by means of screw-bolts.   The connection between the
coils and external leads is made by means of pins passing
through insulating bushes in the circular heads of the
main casting.   The coils are protected by a perforated
covering.   For multiphase work the firm uses a corresponding number of single-phase transformers.

**Messrs. the Elektrizitäts-Aktien-Gesellschaft vormals W.
Lahmeyer and Co., of Frankfort,** use the core type for
simple and multiphase work.   Their designs are shown in
Figs. 114 and 115.   The former is for a single-phase
transformer of 30 Kwt., and the latter for a three-phase
transformer of 40 Kwt., the transforming ratio being in
both cases 5000 : 110 volt.   On account of the high
voltage at the primary, the latter is subdivided into a
large number of sections whereby the danger of a break-
down of the insulation is of course considerably reduced.
The cores have shamfered corners to reduce the length of
wire.   The coils are wound separately on insulating
cylinders.   The other details can easily be gathered from
the diagrams.

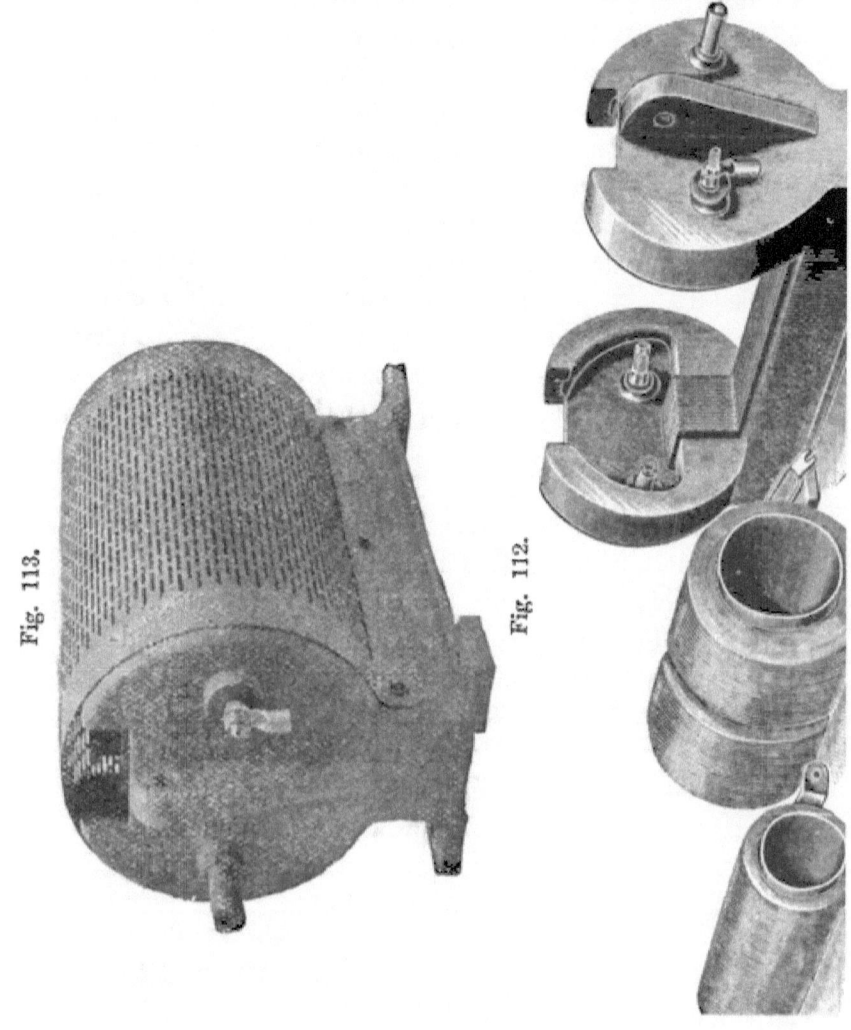

Fig. 113.

Fig. 112.

Fig. 114.

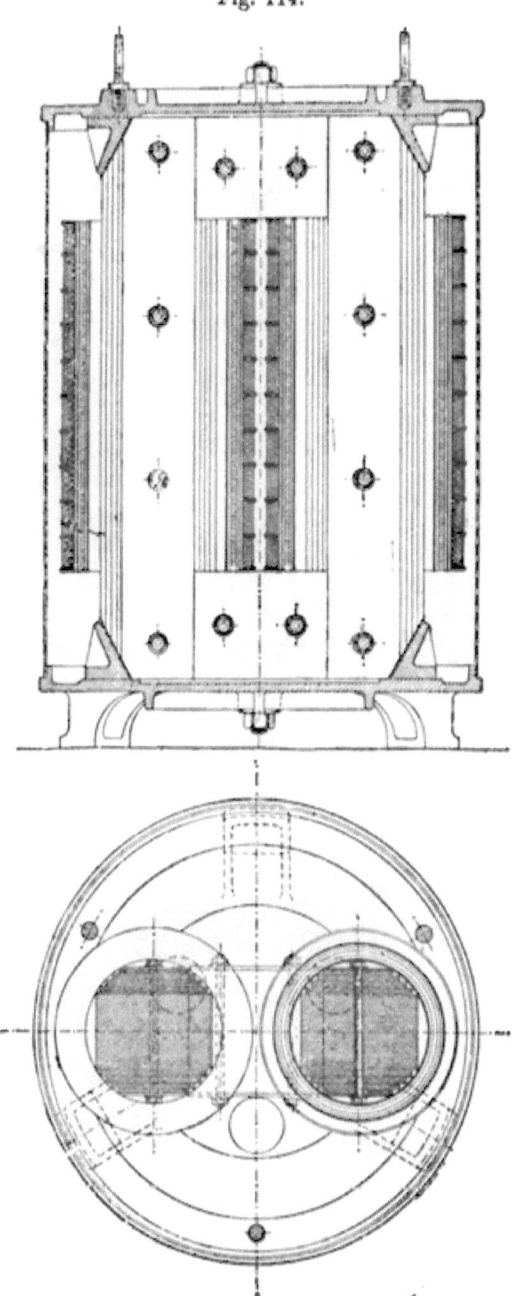

Fig. 115.

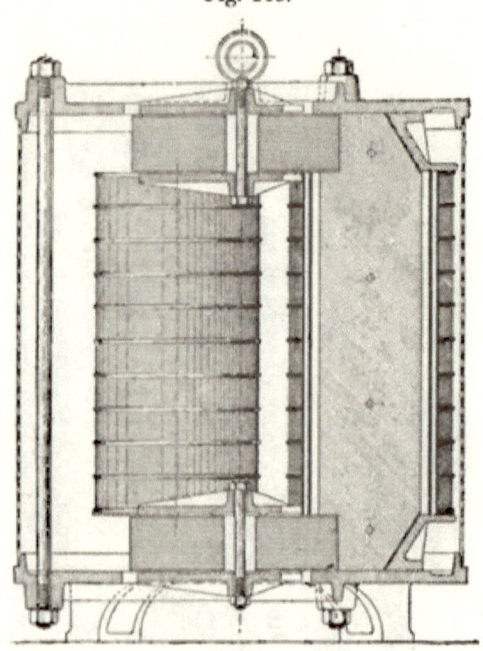

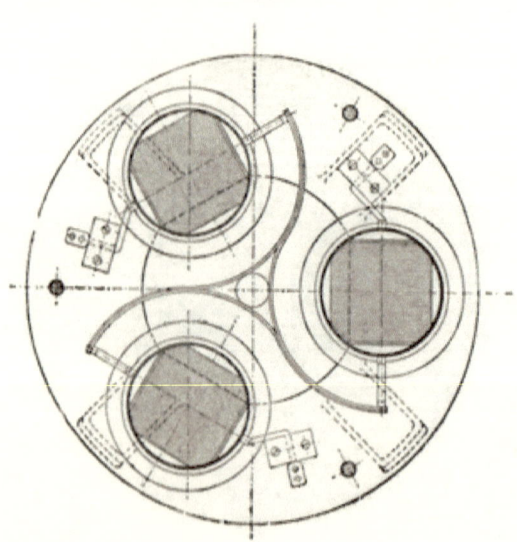

The **Brush Electrical Engineering Company, Limited, of London**, make a shell type of transformer designed by Mr. William Mordey. The manner in which the iron part of the apparatus is built up has already been described in Chapter III. At the ends are strong cast-iron frames, which, by means of screw-bolts, serve for pressing the

Fig. 116.

plates firmly together. (Fig. 116.) The top frame is arranged to take a stout plate of porcelain, on which are placed the terminals, fuses, and a high-pressure double-pole switch. The whole apparatus is then put into a water-tight cast-iron case, as shown in Fig. 117. The spindle of the switch-lever is carried through this case, so that the transformer may be disconnected on the high-pressure side without opening the lid of the case. The

Fig. 117.

connecting cables are brought in through insulated stuffing boxes. According to the makers, these transformers have a drop of $2\frac{1}{2}$ per cent. in all sizes, whilst the hysteresis

loss for $\sim$ = 100 varies from 6 per cent. in the smallest
size for 750 watt, to ¾ per cent. in the largest size of 50
Kwt. The following table gives the weight, including case,
for various sizes.

Fig. 118.

Output Kwt.	1·5	3	6	12	24	50
Weight kgr.	132	247	359	559	863	2083

The two windings are separated by a metallic shield
which is connected to earth, so that the high-pressure
cannot get into the low-pressure winding. (Compare
Chapter VIII.)

Messrs. **Johnson and Phillips, of Charlton**, build core tranformers of the type shown in Figs. 41 to 44. Since these illustrations show all the details, no further description need be given. Fig. 118 shows parts of a 2-Kwt. transformer, and Fig. 119 shows a 10-Kwt. transformer and its case.

Fig. 119.

Messrs. **Ganz and Co., of Budapest**, have adopted the shell type with a short core (Fig. 120). The iron part consists of **E**-shaped stampings, which are held in cast-iron frames of circular contour, so that the whole transformer when finished has the form of a short cylinder, and can be rolled along without being damaged. The terminals are mounted on porcelain washers, and fuses are fitted for the high-pressure winding. The windings are placed on forms of "press-spahn," and are subdivided into sections to reduce magnetic leakage.

Fig. 120.

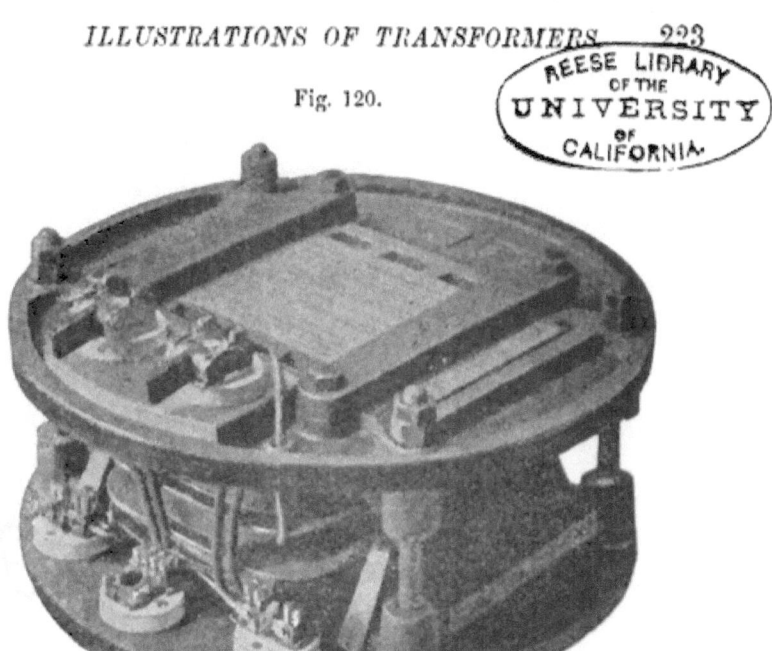

The Maschinenfabrik Oerlikon use the shell type for single phase, and the core type for three-phase transformers. Fig. 121 shows a single-phase transformer. The core is built up of plates of different width (so as to approximate a cylinder in shape), and these are held together by gun-metal side-pieces and bolts. At the ends the cylinder is cut down to the depth necessary to obtain contact with the ⌊____⌋ shaped yokes over their whole width. The coils are separately wound on paper cylinders, care being taken to leave sufficient clearance for pushing one winding over the other. The yoke-plates are driven into corresponding recesses in the cast-iron end-pieces, and are pressed against the case by screw-bolts. To protect the winding perforated covers are placed on both sides, as

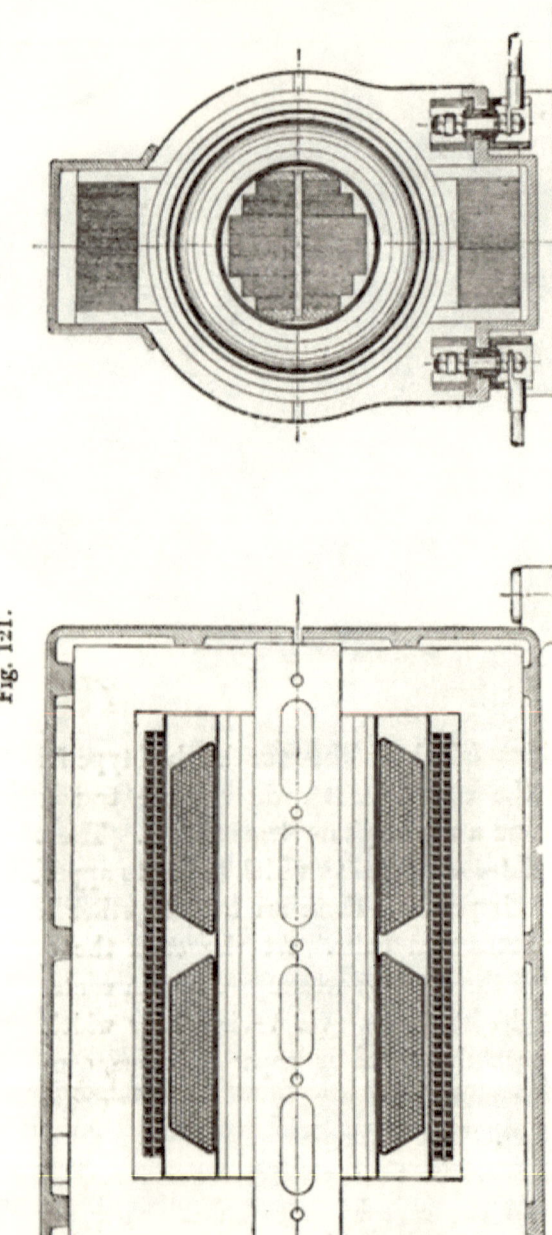

Fig. 121.

shown in Fig. 122. Fig. 123 shows a similar transformer, but without side-covers.

Fig. 122.

Fig. 123.

For three-phase work this firm use two types, the one

Q

symmetrical with circular yokes (Figs. 124 and 125), and the other with straight yokes (Fig. 126). Theoretically, the symmetrical arrangement is preferable, since the length of the magnetic path is the same in each phase; from a practical point of view this advantage is, however, not very important. The magnetic resistance of the butt-

Fig. 124.

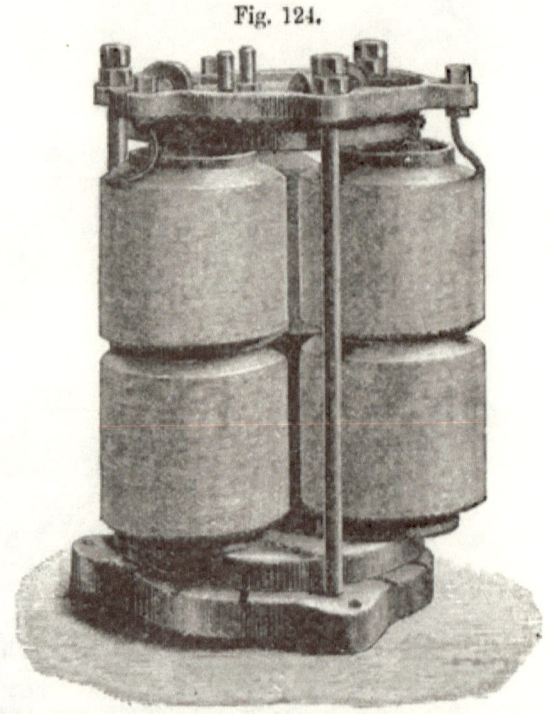

joints is necessarily very much greater than that of the iron itself, and a small difference in the three branches of the latter becomes, therefore, unimportant. On the other hand, the straight-yoke type is more convenient in the manufacture.

The Electric Construction Co., Limited, of Wolverhampton

has been one of the first firms to take up the manufacture
of transformers of the shell type. This type they have
retained, but whereas former designs resembled American

Fig. 125.

patterns, their latest design shows some interesting depart-
ures from the ordinary plans, made with a view to better
utilize the material and increase the efficiency. Fig. 127
shows a 10-Kwt., and Fig. 128 a 40-Kwt. transformer.

The coils have the form of long rectangles, and the iron part is long as compared to the thickness of the core. In the latter apparatus (Fig. 128) an attempt has been made

Fig. 126.

to reduce the weight of iron to the possible minimum by using coils of circular cross-section. What we previously have called the windows are, therefore, not rectangular but circular, and the stampings are circular discs. All these

Fig. 127.

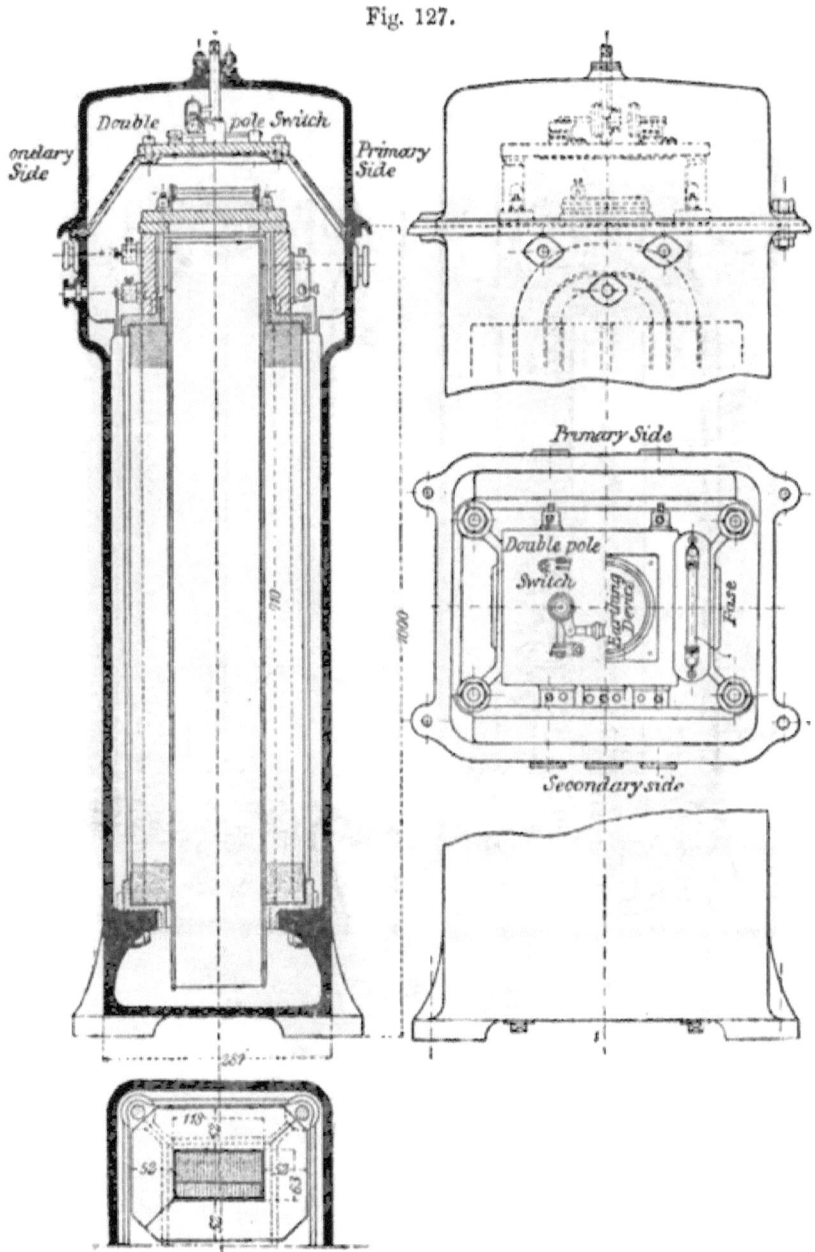

Fig. 128.

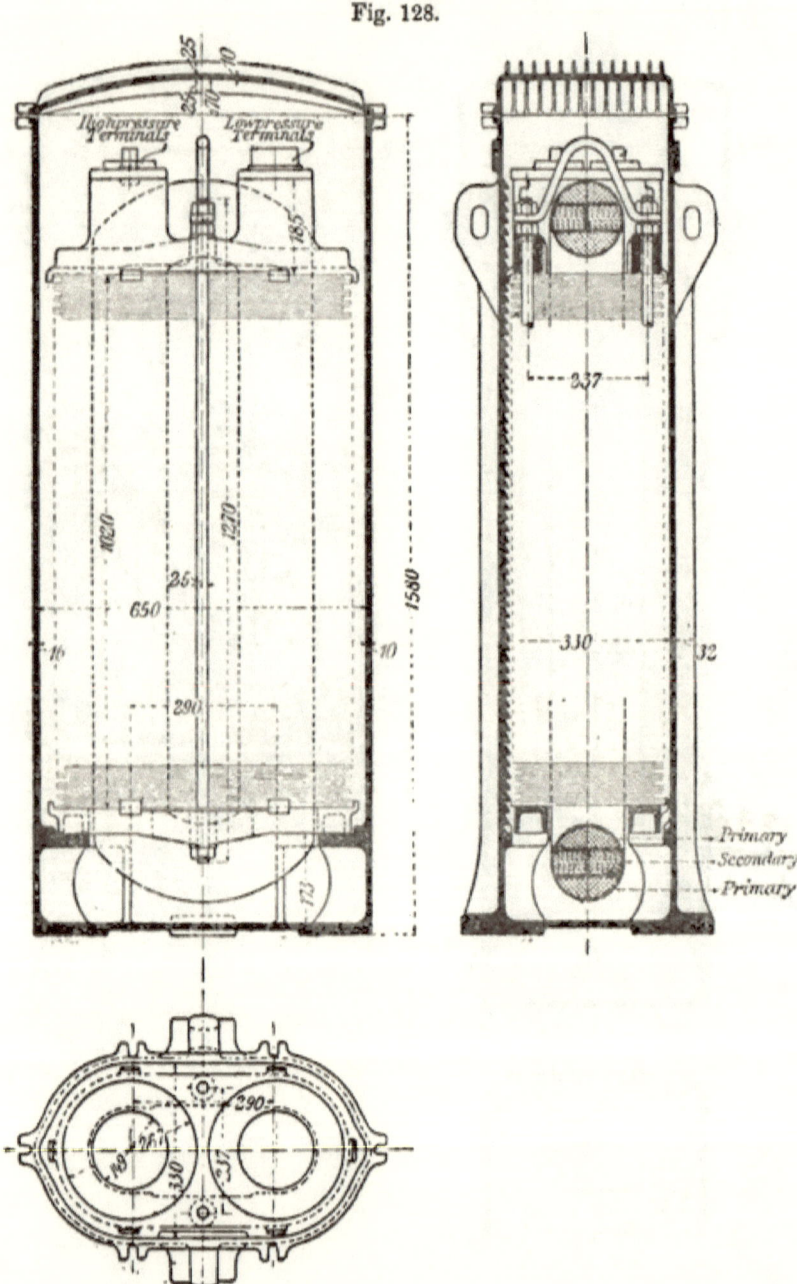

have the same inside diameter, but their outside diameter varies, so that in building up, a parcel of smaller discs is alternated with a parcel of larger discs, by which means the external cooling surface is considerably increased. The capacity of the outer case for taking up heat from the shell of the transformer is increased by the use of

Fig. 129.

corrugations as shown. The case is made long enough to provide room for a double-pole high-pressure switch, double-pole fuses, and a Cardew safety device.

The **Allgemeine Elektricitätsgesellschaft, of Berlin,** has adapted the core type both for single-phase and three-phase transformers. The section of the core is square, with the corners cut back, so as to better fill the circular cavity in the coil, whilst the gun-metal side-plates complete the circular contour. The yokes are similarly provided with side-plates, which serve to hold the iron

Fig. 130.

plates together. The secondary coils are, in the smaller sizes, wound direct on to the heavily insulated core, whereby the length of the secondary winding is reduced

to its possible minimum. The primary coils are wound
on cylinders of micanite. By the employment of this
material the distance between the outer surface of the
secondary and the inner surface of the primary winding
can be reduced to about $\frac{3}{16}''$, so that the magnetic leakage
is exceedingly small. Fig. 129 shows a 10-Kwt. transformer, and two of its coils. A test made with this
transformer, to determine the drop according to the
author's method, as explained in Chapter VI., gave the
following results. The ohmic drop is 2 per cent., whilst
on short circuit 4 per cent. of the primary voltage is
required to produce the full secondary current. The
drop for full load determined from these data is for—

$$\phi = \ 0^\circ \quad \ldots \qquad \ldots \quad 2\cdot3 \text{ per cent.}$$
$$\phi = 60^\circ \quad \ldots \qquad \ldots \quad 4\cdot0 \text{ ,, ,,}$$
$$\phi = 90^\circ \quad \ldots \qquad \ldots \quad 3\cdot9 \text{ ,, ,,}$$

Fig. 130 shows a 40-Kwt. three-phase transformer, of
the type used in the lighting plant at Strassburg. As it
was in this case very important to reduce the drop as
much as possible, the windings are not placed within each
other, but are arranged in small sections side by side.
The voltage on short circuit is in this design only 3 per
cent. of the full voltage, so that the drop does not exceed
3 per cent. even on a motor load. The wire for all the
sections is wound on formers of micanite. Both types
of transformers are protected by perforated side-plates,
not shown in the illustrations.

**The Westinghouse Electric Manufacturing Co., of Pitts-
burg,** are building shell transformers, with short cores and
coils of special shape to facilitate cooling. Fig. 131 shows
the type used for lighting. This is built up to 30 Kwt.
The coils are wound in sections, and where they protrude

from the iron part they are bent out in fan-shape, so as to give a greater cooling surface. The iron part is held together by strong cast-iron frames and bolts, and the

Fig. 131.

coils are protected by a perforated casing of cast-iron. This is partly shown in the illustration. Figs. 132 and 133 show the 100-Kwt. transformers used in the Niagara installation. The transforming ratio is 2000 to 150 volts, but provision is made by means of a second pair of primary terminals to alter this ratio slightly in certain cases. The primary and secondary winding consists each

Fig. 132.

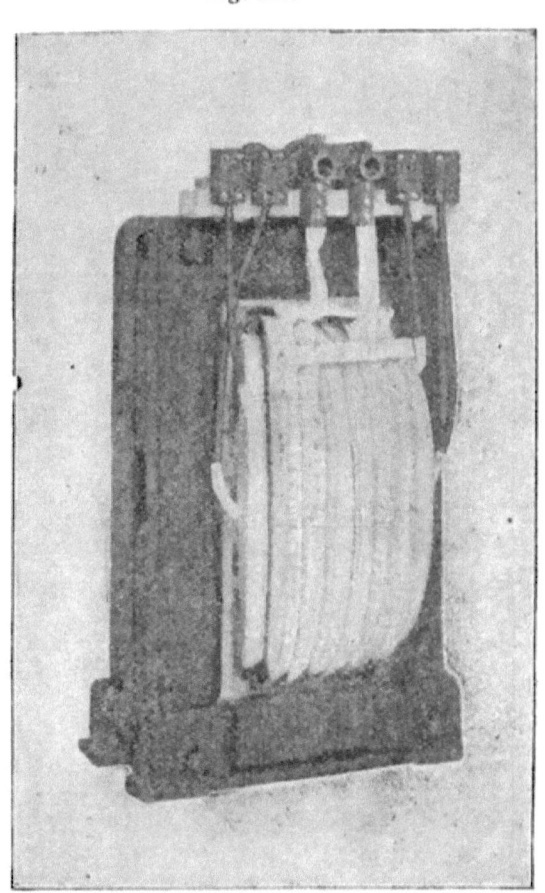

of four sections, the secondary sections being connected in parallel. In the middle of the core are two primary sections, on either side are two secondary sections, and finally a primary section on each outside. The exposed parts of the sections are bent out fan-shaped to increase the cooling surface. The cooling agent is, however, not air, but oil, with which the case of the transformer is filled.

Fig. 133.

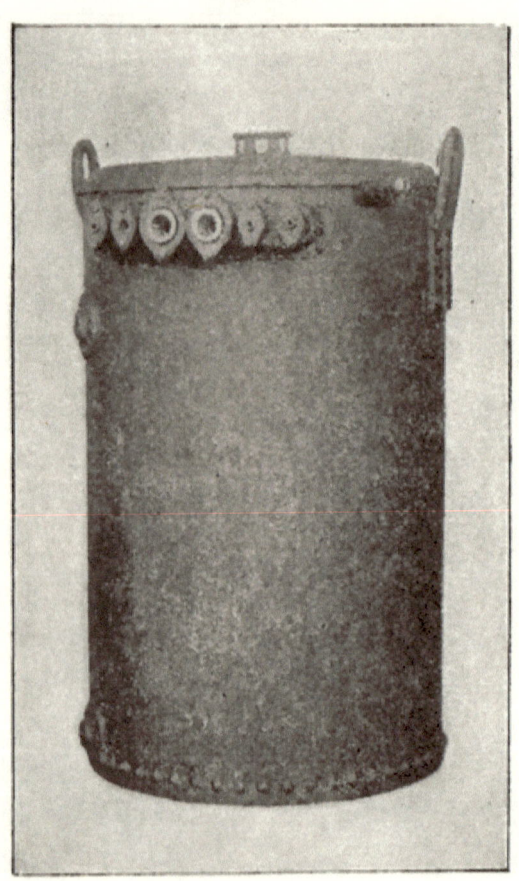

The connecting cables pass through insulated stuffing-boxes as shown in Fig. 133. To cool the oil a wrought-iron spiral tube is fitted to the inner surface of the case, and a stream of cold water is sent through this tube.

INDEX

www.ingramcontent.com/pod-product-compliance
Lightning Source LLC
Chambersburg PA
CBHW020057030726
47498CB00006B/1833